Mining Engineering Analysis

Second Edition

Christopher J. Bise

Published by the
Society for Mining, Metallurgy, and Exploration, Inc.

Society for Mining, Metallurgy, and Exploration, Inc. (SME)
8307 Shaffer Parkway
Littleton, Colorado, USA 80127
(303) 973-9550 / (800) 763-3132
www.smenet.org

SME advances the worldwide mining and minerals community through information exchange and professional development. SME is the world's largest association of minerals professionals.

Copyright © 2003 Society for Mining, Metallurgy, and Exploration, Inc.
Electronic edition published in 2009.

All Rights Reserved. Printed in the United States of America.

Information contained in this work has been obtained by SME, Inc. from sources believed to be reliable. However, neither SME nor its authors guarantee the accuracy or completeness of any information published herein, and neither SME nor its authors shall be responsible for any errors, omissions, or damages arising out of use of this information. This work is published with the understanding that SME and its authors are supplying information but are not attempting to render engineering or other professional services.

No part of this publication may be reproduced, stored in a retrieval system, or transmitted in any form or by any means, electronic, mechanical, photocopying, recording, or otherwise, without the prior written permission of the publisher. Any statements or views presented here are those of the authors and are not necessarily those of SME. The mention of trade names for commercial products does not imply the approval or endorsement of SME.

ISBN 0-87335-221-1

Contents

LIST OF FIGURES vii

LIST OF TABLES ix

PREFACE TO THE SECOND EDITION xi

PREFACE TO THE FIRST EDITION xiii

CHAPTER 1 **PRINCIPLES OF MINE PLANNING** 1
International Standardization 2
Fundamentals 5
Problems 9
Problem Solutions 13

CHAPTER 2 **MINE PREPLANNING** 19
Reserve Estimation 19
Production Planning 22
Staff Planning 24
Project Scheduling 26
Mine Safety and Health 27
Quality Control 31
Problems 34
Problem Solutions 41

CHAPTER 3 **VENTILATION** 51
Fundamentals of Airflow 52
Fan Selection 56
Ventilation Requirements for a New Mine 58
Problems 59
Problem Solutions 66

CHAPTER 4 **STRATA CONTROL** 79
Stresses Around Mine Openings 79
Safety Factor 82

                    Pillar Design   82
                    Roof-Span Design   86
                    Selection of Longwall Shields   88
                    Subsidence   88
                    Slope Stability   94
                    Problems   96
                    Problem Solutions   102

CHAPTER 5   **PUMPING AND DRAINAGE**   121
                    Pump Characteristic Curves   121
                    Pipe Characteristic Curves   122
                    Pump Flexibility   123
                    Pumping Applications   125
                    Problems   127
                    Problem Solutions   131

CHAPTER 6   **MINE POWER SYSTEMS**   145
                    Compressed-Air Power   145
                    Electrical Power   153
                    Problems   165
                    Problem Solutions   172

CHAPTER 7   **HOISTING SYSTEMS**   195
                    Wire Ropes   195
                    Hoists   197
                    Problems   204
                    Problem Solutions   206

CHAPTER 8   **RAIL AND BELT HAULAGE SYSTEMS**   215
                    Track Layouts   215
                    Locomotive Tractive-Effort Calculations   217
                    Locomotive Motor Duty Cycle   219
                    Belt Conveyors   221
                    Problems   227
                    Problem Solutions   230

CHAPTER 9   **RUBBER-TIRED HAULAGE SYSTEMS**   241
                    Underground Face-Haulage Vehicles   241
                    Surface Haulage Vehicles   245
                    Problems   251
                    Problem Solutions   255

CHAPTER 10   **SURFACE EXTRACTION**   **271**
Fragmentation Procedures   **271**
Design of Blasting Rounds   **273**
Number of Drills   **279**
Amount of Explosives Required   **280**
Excavation Fundamentals   **280**
Dragline Selection   **280**
Shovel Selection   **284**
Bulldozer Selection   **284**
Scraper Selection   **286**
Front-End Loader Selection   **287**
Problems   **288**
Problem Solutions   **296**

INDEX   **307**

ABOUT THE AUTHOR   **313**

Figures

1.1	Diagram for mine planning	**2**
2.1	Method of polygons (plan and polygonal prisms)	**20**
2.2	Construction of perpendicular bisectors	**21**
2.3	Correct construction of polygons by perpendicular bisectors	**21**
2.4	Incorrect construction of polygons	**22**
2.5	Square net pattern	**22**
2.6	Chessboard pattern	**23**
2.7	PERT program network	**27**
2.8	Examples of control charts	**32**
2.9	Normal distribution showing range of standard deviation	**33**
3.1	Single, double, and triple airways	**54**
3.2	Splitting of airflow	**55**
3.3	Relationship between fan characteristic and mine resistance	**57**
3.4	Typical line diagram for a mine ventilation system	**59**
4.1	Original stress distribution on a virgin coal seam	**80**
4.2	Stress trajectories around a single opening	**80**
4.3	Boundary stress concentration for a circular hole in a biaxial stress field	**82**
4.4	Boundary stress concentration for ovaloidal holes in a biaxial stress field	**83**
4.5	Boundary stress concentration for rectangular holes with rounded corners	**83**
4.6	Critical compressive stress concentrations around openings with various shapes	**84**
4.7	Dimensions for the determination of shield ratings	**89**
4.8	Subsidence development curve	**90**
4.9	Graph for predicting maximum slope and strain for various width-to-depth ratios of a panel	**91**
4.10	Relationship of damage to length of structure and horizontal ground strain	**92**
4.11	Protection practice of surface structures in Pennsylvania	**93**
4.12	Planar failure	**94**
5.1	Centrifugal pump testing arrangement	**122**
5.2	Typical characteristic curves for a centrifugal pump	**122**
5.3	Components of head	**125**
5.4	Combined characteristic for two pumps in series	**126**

6.1	The various steps in a reciprocating compressor cycle	**147**
6.2	Combined theoretical indicator card for a two-stage two-element 100-psig positive-displacement compressor	**148**
6.3	Adiabatic and isothermal compression chart	**148**
6.4	Effect of cylinder clearance on volumetric efficiency	**149**
6.5	A radially distributed underground power system	**154**
6.6	Utilization in a continuous-miner system	**154**
6.7	Power factor triangles	**161**
7.1	Areas of applications of drum or Koepe hoists	**198**
7.2	Minimum factors of safety for wire rope	**199**
7.3	Suggested drum diameter (D)–rope diameter (d) ratio	**199**
7.4	Approximate equivalent effective weight reduced to rope center for different diameter drums	**200**
7.5	Horsepower per time cycle for a drum hoist	**201**
7.6	Horsepower per time cycle for a friction hoist or a drum hoist with a tail rope	**203**
8.1	Typical rail turnout with components indicated	**216**
8.2	Variables in turnout calculations	**216**
8.3	Relationship between a locomotive and its trailing load	**218**
8.4	Conveyor belt arrangements	**224**
9.1	Sample cut sequence in an underground coal mine	**242**
9.2	Haulage vehicles at the intersection containing the change point	**243**
9.3	Typical rim-pull-speed-gradability curves for an off-highway truck	**246**
9.4	Travel time (loaded) for a typical off-highway truck	**247**
9.5	Travel time (empty) for a typical off-highway truck	**248**
9.6	Typical brake performance (continuous grade retarding) for an off-highway truck	**249**
9.7	Typical brake performance (1,500 ft) for an off-highway truck	**249**
9.8	Typical brake performance (2,000 ft) for an off-highway truck	**250**
9.9	Typical brake performance (3,000 ft) for an off-highway truck	**250**
9.10	Typical brake performance (5,000 ft) for an off-highway truck	**251**
10.1	Nomograph for determining loading density	**273**
10.2	Nomograph for determining detonation pressure	**273**
10.3	Parameters of a blasting round	**274**
10.4	Federal standards for blast vibration	**275**
10.5	Bench cross section view showing D_e, B, H, J, T, and L	**275**
10.6	Generalized blasting patterns showing B, S, b, and s	**276**
10.7	Dragline pit	**283**
10.8	Dragline pit with a working bench	**284**
10.9	Shovel pit without a coal fender	**285**
10.10	Loaded scraper (distance vs. time)	**286**
10.11	Empty scraper (distance vs. time)	**287**

Tables

1.1	SME abbreviated conversion table	**3**
2.1	Computation of reserves by the polygonal method	**20**
2.2	Typical staffing for underground coal room-and-pillar and longwall mines	**24**
2.3	Subparts in 30 CFR that pertain to mandatory safety and health standards in underground coal mines	**29**
2.4	Subparts in 30 CFR that pertain to mandatory safety and health standards in surface coal mines and surface work areas of underground coal mines	**30**
2.5	Values for combining decibel levels	**31**
2.6	Permissible noise exposures	**31**
3.1	Airflow principles	**52**
3.2	Bureau of Mines schedule for friction factors of mine airways	**53**
3.3	Friction factors (K) for coal mine airways and openings	**54**
3.4	Fan laws	**56**
4.1	Some properties of rock and coal materials	**81**
4.2	Relationship between W/D and d/D for various points on a subsidence profile	**91**
4.3	Relationship for various strain values in a subsidence profile	**92**
5.1	Equivalent number of feet of straight pipe for different fittings	**123**
5.2	Friction loss, in feet, for old pipe ($C = 100$)	**124**
6.1	Air consumption multipliers for altitude operation of rock drills based on 80 to 100 psig air pressure	**150**
6.2	Multipliers for air consumption of rock drills	**150**
6.3	Factors for correcting the actual capacity of single-stage compressors at sea level when they are used at higher altitudes	**150**
6.4	Atmospheric pressure at different altitudes	**151**
6.5	Air requirements of representative drilling machines	**151**
6.6	Pressure loss in hose	**152**
6.7	Ampacities for portable power cables in amperes per conductor	**156**
6.8	Ampacities for three-conductor mine power cables	**157**

6.9	Correction factors for insulations rated at 90°C for various ambient temperatures	**157**
6.10	Correction factors for cables used with one or more layers wound on a reel	**157**
6.11	Resistance and reactance of portable power cables	**158**
6.12	Resistance and reactance of mine power feeder cable	**159**
6.13	Power factors of typical a-c loads	**160**
6.14	kVAR table	**161**
6.15	Examples of default data for mining equipment	**163**
6.16	Typical power billing schedule	**164**
7.1	Specifications for 6 × 19-class haulage ropes	**196**
7.2	Specifications for 6 × 7-class haulage ropes	**196**
7.3	Minimum factors of safety (*SF*) for wire ropes	**196**
8.1	Suggested rail weights	**216**
8.2	Adhesive values for locomotives	**218**
8.3	Maximum recommended lump size for various belt widths	**222**
8.4	Typical maximum belt speeds in feet per minute	**222**
8.5	Normal bulk material capacity of troughed conveyor belts in tons per hour for each 100 fpm of belt speed	**223**
8.6	Friction and length factors for conveyor belts	**225**
8.7	B and Q for conveyor belts	**225**
8.8	K values for conveyor belts	**226**
8.9	Idler spacing for conveyor belts	**226**
9.1	Industry-accepted standards of rolling resistance factors	**245**
9.2	Approximate coefficient of traction factors	**247**
10.1	Worksheet for design of blasting rounds	**278**
10.2	Bucket-capacity parameters	**281**
10.3	Shovel operating efficiency	**281**
10.4	Bank density, swell factor, and fillability of common materials	**282**
10.5	Shovel swing factor	**283**
10.6	Propel time factor	**283**

Preface to the Second Edition

A little more than 15 years ago, when I wrote the first edition of this text, my desire was to create a book that would fill the gap between theory and application. Each chapter was organized to show students and industry professionals how to apply various equations to solve typical problems. It was also written to provide a link to widely used but out-of-print textbooks. The second edition continues with these concepts in mind.

The first edition provided practice problems with solutions. As a result, individuals preparing for the Professional Engineers' Examination in Mining Engineering found it useful as a study guide. This edition has been updated to provide new information, without significantly altering the organization of the book. However, one major change in the book's organization has been to place all problem solutions at the end of each chapter. This was done so that people preparing for the P.E. Exam would find it easier to use the text as a drill-and-practice study guide.

Many friends have been instrumental in the completion of this work. I would like to thank my colleagues at Penn State for their suggestions, particularly Marek Mrugala and L. Barry Phelps. I would also like to extend my appreciation to Barry Miller and Philip Melick, two of my former students; Harry Martin of DBT America; and Neal L. Locke of Allegheny Power who provided information that was incorporated into this edition. Also, I would like to thank Madan M. Singh, president of Engineers International, Inc., for his valuable review of the strata control chapter. Finally, I would like to express my sincere appreciation to Jane Olivier, Manager of Book Publishing for SME, for her encouragement and support.

Dr. Christopher J. Bise

Professor and Program Chair of Mining Engineering and Industrial Health and Safety

George H., Jr., and Anne B. Deike Chair in Mining Engineering

The Pennsylvania State University
University Park, PA

Preface to the First Edition

Over the past several years, it has become readily apparent to the members of the mining engineering profession that our field lacks a suitable number of textbooks for teaching and reference. This problem is magnified when a comparison is made with the number of texts available in other engineering disciplines. Another problem has been that most recent mining engineering books tend to deal solely with theory; very few of them show applications of the theory. Unfortunately, the people using the books—students and professionals in the field—want to know how to apply the theory to everyday problems. It should come as no surprise, therefore, that many practicing mining engineers treasure their copies of texts written over forty years ago because they display more balance between theory and applications. In fact, many college mining courses are based on professors' lecture notes rather than published textbooks.

This book is an attempt to fill the gap between theory and application. It was written with two audiences in mind: (1) the mining engineering student who is studying mine design and requires guidance in assembling a mine-design project, and (2) the industry professional who requires a mine-design reference book for daily use or, perhaps, to prepare for the professional engineers' examination. It was written as a complement to two SME books: *SME Mining Engineering Handbook* and *Coal Mining Technology-Theory and Practice*. The problems were based upon course material taught at Penn State and the P.E. Review short course sponsored by Penn State and SME. Admittedly, no text of this nature can contain problems covering all aspects of the discipline, but the construction of the book is designed to allow future update without making the initial text obsolete. Also, more emphasis is given to those subjects that lack suitable modern textbooks as opposed to subjects, such as ventilation, that have several up-to-date texts.

Many friends have been instrumental in the completion of this work. I would like to thank Prof. Robert L. Frantz, Dr. Raja V. Ramani, and my other colleagues in the Department of Mineral Engineering at Penn State for their support. I would particularly like to thank Joan Andrews for typing most of the problems and solutions although, at one time or another, every secretary in the department aided in the book's preparation and has my gratitude. No matter how carefully a book is prepared, proofreading is essential. Therefore, I would like to extend my appreciation to five of my former students—Michael Moore, John Weiss, Joseph Sottile, David Williamson, and Ronald Hetzer—and also to Judy Kiusalaas, Editor for Penn State's College of Earth and Mineral Sciences, for their efforts in this endeavor. Others who have reviewed individual chapters are: Dr. Lawrence Adler (West Virginia University), Dr. Lloyd Morley (The University of Alabama), Dr. Erdal Unal (Middle East Technical University, Ankara,

Turkey), Andrew Drebitko (Drew Development Co.), William Brown (Rexnord Process Machinery Division), and A. R. Worster (Ingersoll-Rand Company). Finally, I would like to express my appreciation to Marianne Snedeker, Manager of Publications for SME, for her efforts in preparing the text.

Dr. Christopher J. Bise

Associate Professor of Mining Engineering

The Pennsylvania State University
University Park, PA
December, 1985

CHAPTER 1

Principles of Mine Planning

The process of planning a mine can be reduced to a network of interrelated systems that are tied together by a common philosophy of mine planning: namely, that the resource being mined is to be extracted in a safe, efficient, and profitable manner. The success of any mining operation cannot be guaranteed unless each of these three requirements is met. Consequently, during the initial planning stages, all mining operations are analyzed and evaluated in a similar fashion.

Figure 1.1 illustrates a diagram for the initial planning stages of any mining operation. Notice that a preplanning stage leads to the planning of four interrelated systems. This preplanning stage includes property exploration; estimation of production and staffing requirements; recognition of legal, environmental, and health and safety standards; and establishment of construction timetables. The four systems evaluated after the preplanning stage are further reduced to the: (1) excavation and handling system, (2) strata control system, (3) operating support system, and (4) service support system.

The excavation and handling system includes selecting all the equipment needed to remove the resource from its in-place deposit (e.g., draglines, continuous miners, rock drills, etc.) and to transport the mined material from the mine face (e.g., belt conveyors, trucks, etc.). The strata control system incorporates everything to be considered in order to maintain the integrity of the roof, ribs, floor, and overburden, both during and after the extraction phase. Typical considerations include pillar sizing, artificial support of the immediate roof in mine entries, and subsidence protection. The operating support system contains the auxiliary requirements of mine planning; this system consumes a large percentage of the mining engineer's planning time but does not contribute to the actual extraction of the resource. However, if proper attention is not paid to the various components of the operating support system, such as ventilation, drainage, and power, the mining operation will not perform to its capabilities. Finally, the service support system is composed of those operations that keep the mine running efficiently; typical examples include maintenance and supply. One important concept that Figure 1.1 is designed to emphasize is that the four systems are interrelated; often, a slight alteration of one can affect the performance of another. If these interrelationships are ignored, the effects on the overall project can be disastrous.

The mine preplanning stage and the four subsequent system evaluations constitute the analysis portion of the mine-planning diagram. After this portion of the study is completed, the overall projected operation can be evaluated using techniques such as financial analysis. This last step in the study is referred to as the synthesis of the previously calculated parameters into an overall plan.

2 | MINING ENGINEERING ANALYSIS

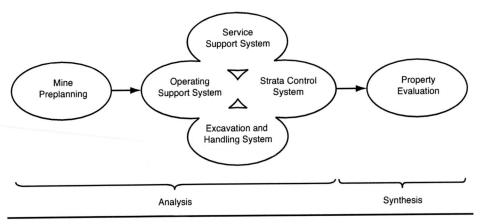

FIGURE 1.1 Diagram for mine planning

Much has been written about the overall evaluation of mining properties; however, only recently has anything been written concerning the coordinated analysis of mine-planning systems. Further, many of the texts that have attempted to address this topic fail give sufficient examples for successful implementation, thus supporting the operating engineers' lament that it is easy to discuss theories but difficult to provide applications of those very same theories.

In keeping with the philosophy behind the first edition of this text, this second edition was written to provide additional applications. Examples of the various stages of the mine-planning flowchart shown in Figure 1.1 are covered in this text, with the exception of the service support system, because this system is oriented more toward day-to-day planning than long-term planning.

INTERNATIONAL STANDARDIZATION

There are numerous benefits to using internationally standardized units for mining engineering calculations. The interchangeability of equipment parts and the ability to analyze and compare performance and design specifications on an equal basis are just two examples of why standardization is important, particularly for manufacturers and mine operators. Unfortunately, engineers in the United States have not rapidly embraced conversion to an international standard; this hestiation is, in part, due to training and familiarity with the English system.

The Society for Mining, Metallurgy, and Exploration, Inc. has committed to adoption of the Systeme International d'Unites (SI) in all mining engineering calculations. However, because many practicing mining engineers and students in undergraduate programs still find it difficult to use SI units, the calculations in this textbook use the English system. Table 1.1 can be used to convert to SI units.

Terminology

The field of mining has its own unique vocabulary, which assigns specific meanings to operations, locations, and functions found in mines. These words, when used in technical writing, can often be confusing to someone unfamiliar with the field. To minimize

TABLE 1.1 SME abbreviated conversion table

Convert from	To	Multiply by
acre	meter squared (m²)	4046.856
acre-foot	meter cubed (m³)	1233.482
ampere-hour	coulomb (C)	3600.00
atmosphere (normal)	kilopascal (kPa)	101.325
barrel (petroleum, 42 gal)	meter cubed (m³)	0.158 99
British thermal unit (Int'l Table)	kilojoule (kJ)	1.055 056
Btu (Int'l Table) per pound (mass)	kilojoule/kilogram (kJ/kg)	2.326
Btu (Int'l Table) per hour	watt (W)	0.293 07
Btu per short ton	kilojoule/ton (kJ/t)	1.163 00
calorie (Int'l Table)	joule (J)	4.186
degree (angle)	radian (rad)	0.017 45
degree Celsius	kelvin (K)	$t_K = t_C + 273.15$
degree Fahrenheit	degree Celsius (C)	$t_C = (t_F - 32)/1.8$
degree Fahrenheit	kelvin (K)	$t_K = (t_F + 459.67)/1.8$
dyne	newton (N)	0.000 01
electron volt	attojoule (aJ)	0.160 219
erg	microjoule (μJ)	0.100
foot	meter (m)	0.3048
foot, cubic	meter cubed (m³)	0.028 316 8
foot, square	meter squared (m²)	0.092 903 04
foot per hour	millimeter/sec (mm/s)	0.084 666 7
foot per minute	meter/sec (m/s)	0.005 080
foot per second	meter/sec (m/s)	0.304 800
foot per sec per sec	meter/sec/sec (m/s²)	0.304 800
gallon (US liquid)	meter cubed (m³)	0.003 785 4
gallon (US liquid)	liter (L)	3.785 412
horsepower (550 ft-lbf per sec)	kilowatt (kW)	0.745 699
inch	millimeter (mm)	25.400
inch, square	millimeter squared (mm²)	645.160
inch, cubic	millimeter cubed (mm³)	16 387.06
inch per second	millimeter/sec (mm/s)	25.40
inch of mercury (60°F)	kilopascal (kPa)	3.376 85
inch of water (60°F)	kilopascal (kPa)	0.248 84
kilocalorie (Int'l Table)	kilojoule (kJ)	4.186 80
kilowatt-hour	kilojoule (kJ)	3600.00
mile (US statute)	kilometer (km)	1.609 344
mile, square (US statute)	kilometer squared (km²)	2.589 988
mile per hour (US statute)	kilometer/hour (km/h)	1.609 344
millimeter of mercury (0°C)	pascal (Pa)	133.3224
minute (angle)	radian (rad)	0.000 291
ohm (Int'l US)	ohm (Ω)	1.000 495

(table continues on next page)

TABLE 1.1 SME abbreviated conversion table (continued)

Convert from	To	Multiply by
ounce-mass (avoirdupois)	gram (g)	28.349 520
ounce-mass (troy)	gram (g)	31.103 480
ounce (US fluid)	centimeter cubed (cm^3)	29.573 530
pound-force (lbf avoirdupois)	newton (N)	4.448 222
pound-force per square inch (psi)	kilopascal (kPa)	6.894 757
pound-mass (lbm avoirdupois)	kilogram (kg)	0.453 592 4
pound-mass per short ton	kilogram/ton (kg/t)	0.500 000
quart (US liquid)	liter (L)	0.946 325 4
ton (long, 2,240 lbm)	ton (t)	1.016 047
ton (short, 2,000 lbm)	ton (t)	0.907 184
ton (metric)	kilogram (kg)	1000.0
watt-hour	kilojoule (kJ)	3.600
yard	meter (m)	0.9144
yard, square	meter squared (m^2)	0.836 127 4
yard, cubic	meter cubed (m^3)	0.764 544 9

Source: Dasher 1981.

confusion, this section defines many of the terms used in the text. A more extensive dictionary of mining terminology is the American Geological Institute's *Dictionary of Mining, Mineral, and Related Terms*, available from SME and cited at the end of this chapter.

Advance mining: Exploitation in the same direction, or order of sequence, as development.

Air course: A passage through which air is circulated.

Air split: The division of a current of air into two or more parts.

Airway: Any passage through which air is carried.

Barrier pillars: Solid blocks of coal or rock left between two mines or sections of a mine to prevent accidents caused by inrushes of water or gas or to protect pillars in headings against crushing.

Beam building: The creation of a strong, inflexible beam by bolting or otherwise fastening together several weaker layers. In coal mining, this is the intent of roof bolting.

Belt idler: A roller, usually of cylindrical shape, that is supported on a frame and that, in turn, supports or guides a conveyor belt.

Belt take-up: A belt pulley, generally under a conveyor belt and usually located near the drive pulley, kept under strong tension parallel to the belt line. Its purpose is to automatically compensate for any slack in the belting created by start-up, etc.

Bleeder entries: Special air courses developed and maintained as part of the mine ventilation system and designed to continuously move air–methane mixtures emitted by the gob away from active workings and into mine-return air courses.

Breakthrough: A passage for ventilation that is cut through the pillars between rooms.

British thermal unit (Btu): Heat needed to raise 1 pound of water 1 degree Fahrenheit (1°F).

Burden (of holes in blasting): The distance between the rows of holes running parallel to the free vertical surface of the rock.

Cage: A device in a mine shaft, similar to an elevator car, that is used to hoist men and materials.
Car: A railway wagon, especially any of the wagons adapted to carrying coal, ore, and waste underground.
Collar: The timbering or concrete around the mouth or top of a shaft.
Cover: The overburden above any deposit.
Crosscut: A passageway driven between the entry and its parallel air course or air courses for ventilation purposes.
Development: The work done on a mine after exploration to provide access to the ore and to provide haulage ways for the exploitation period.
Dump: The point at which a load of ore or coal is discharged.
Entry: An underground passage used for haulage or ventilation.
Face: The principal operating place in a mine.
Gob: That part of the mine from which the coal has been removed and the space has been filled up with waste rock.
Head: Pressure of a water column in feet.
Headframe: The structure surmounting the shaft that supports the hoist rope pulley and, often, the hoist itself.
Inby: In the direction of the working face.
Intake: The passage through which fresh air is drawn or forced into a mine or into a section of a mine.
Main entry: A main haulage road.
Outby: In the direction of the mine entrance.
Panel: A coal mining block that generally comprises one operating unit.
Portal: The structure surrounding the immediate entrance to a mine.
Retreat mining: Exploitation in the direction opposite from development.
Return: The air or ventilation that has passed through all the working faces of a split.
Rubbing surface: The total area around an airway.
Skip: A car being hoisted from a slope or shaft.
Spacing (of holes in blasting): The distance between two holes in the same row.
Split: Any division or branch of the ventilation circuit.
Subsidence: The deformation of the ground mass surrounding a mine due to the mining activity.
Trip: A train of mine cars.

FUNDAMENTALS

Because most calculations used in mining engineering practice originated in other science and engineering disciplines, a brief review of science and engineering fundamentals is imperative at the outset. Thus, these fundamentals are defined in the following section so that a review will not be necessary in later chapters.

Displacement: The distance covered by an object from a fixed origin.
Velocity: The ratio of an object's displacement to the time interval during which displacement occurred.

Acceleration: The change in velocity during a particular time interval.

Relationship among displacement, velocity, and acceleration: The distance, s, that an object travels during a particular time period, t, is related to the average velocity, v_{avg}, in the following manner:

$$s = v_{avg} t \quad \text{(EQ 1.1)}$$

If the body's velocity changes at a uniform rate from an initial velocity, v_i, to a final velocity, v_f, the body is said to have a constant acceleration, a.

$$a = (v_f - v_i) / t \quad \text{(EQ 1.2)}$$

$$v_f = v_i + at \quad \text{(EQ 1.3)}$$

Because the acceleration is constant, the average velocity of the object is

$$v_{avg} = 0.5(v_i + v_f) \quad \text{(EQ 1.4)}$$

The displacement becomes

$$s = 0.5(v_i + v_f)t = v_i t + 0.5(at^2) \quad \text{(EQ 1.5)}$$

The final velocity becomes

$$v_f^2 = v_i^2 + 2as \quad \text{(EQ 1.6)}$$

Force: An action that changes or tends to change the state of rest or motion of a body. It is related to displacement, velocity, and acceleration in the following manner:

$$f = ma = (W/g)\, a = (W/g)\, (v_f - v_i) / t \quad \text{(EQ 1.7)}$$

where g is the acceleration of a freely falling body and has a value of 32.2 fpsps (feet per second per second), m is the mass of the body, and W is the weight of the body.

Work: The product of a force unit and a length unit in which a force does work when it acts against a resisting force to produce motion of an object. For example, 1 ft-lb of work is done when a constant force of 1 lb moves a body a distance of 1 ft in the direction of the force.

Power: The time rate of doing work. Average power is defined as either the quotient of the amount of work done and the time taken to do the work or the product of the force applied and the velocity of the object in the direction of the applied force. The two units of power used in most mining situations are the watt (W) and horsepower (hp). They are related to each other in the following manner:

$$1 \text{ hp} = 33{,}000 \text{ ft-lb per min} = 746 \text{ W}$$

Specific gravity (s.g.): The ratio of the weight of a substance to the weight of an equal volume of a standard substance. Solids and liquids are compared to a standard of water; gases are often compared to a standard of air.

Efficiency: The ratio of work output to work input, or power output to power input.

Recovery ratio: In mining, a special measure of efficiency. It is defined as the volume of mineral mined from a deposit divided by the total volume of mineral in the deposit prior to mining.

Stress: The force applied per unit area that produces or tends to produce deformation in a body.

Strain: A pure, dimensionless number that represents the fractional deformation resulting from a stress. For example, if a rock specimen of an initial length experiences an elongation due to a tension test, the longitudinal strain is

$$\text{(change in length)/(initial length)} \quad \text{(EQ 1.8)}$$

Modulus of elasticity: The constant that reflects the fact that when a body tends to return to its original size or shape after a deformation or when the deforming forces have been removed, it is said to have the property of elasticity, and within the elastic limit of any body, the ratio of the stress to the strain produced is constant.

Quantity of a fluid (or gas) flowing through a duct: This quantity is equal to the product of the fluid's velocity and the cross-sectional area of the duct.

Swell factor: The ratio of the bank value to the loose value when a material has been disturbed by an excavation process and loses volume as opposed to the in-place or bank volume.

Rolling resistance (of a vehicle): The sum of the external forces opposing motion over level terrain.

Grade resistance (assistance): The resistance (or assistance) to motion of a vehicle that is caused by the pull of gravity as it travels over grades.

Tractive effort: The effort exerted by a prime mover at the rim of its driving wheels. It is a function of the prime mover's weight and its ability to adhere to the roadbed.

Drawbar pull: That portion of the prime mover's tractive effort that is available to move a trailing load after the resistances of the prime mover are subtracted.

Ohm's law: A fundamental law of circuit analysis that stipulates that the force necessary to impart motion on a fluid through a conductor is equal to the product of the volume flowing and the resistance to flow. Electrically, Ohm's law states that voltage is equal to the product of current and resistance.

Kirchhoff's laws:

1. The algebraic sum of the fluid volume flowing into a junction is zero. Electrically, this fluid volume is current.
2. The algebraic sum of pressure drops around a closed circuit is zero. Electrically, this pressure is voltage.

Kinetic energy: An object's ability to do work because of its motion. The kinetic energy, KE, of a mass, m, moving with a velocity, v, is:

$$KE = 0.5(mv^2) \quad \text{(EQ 1.9)}$$

Conservation of energy: The fact that energy can neither be created nor destroyed; it can only be transformed from one form to another.

General gas laws: At sufficiently low pressures and high temperatures, all gases have been found to obey the following three laws, which relate the volume of a gas to the pressure and temperature: Boyle's law, Charles's law, and Gay-Lussac's law. Any two of the three gas laws may be combined to obtain

$$\frac{pV}{T} = \text{constant} \quad \text{(EQ 1.10)}$$

Further, if a given mass of gas is considered under two different conditions of temperature and pressure, the following holds true:

$$\frac{p_1 V_1}{T_1} = \frac{p_2 V_2}{T_2} \qquad \text{(EQ 1.11)}$$

Bernoulli theorem: The equation that results from application of the principle of conservation of energy to fluid flow. The energy possessed by a flowing fluid consists of internal energy and energies due to pressure, velocity, and position. In the direction of flow, the energy principle is described by the following equation for steady flow of incompressible fluids in which the change in internal energy is negligible:

$$\left(\frac{p_1}{w} + \frac{V_1^2}{2g} + z_1\right) + H_a - H_l - H_e = \frac{p_2}{w} + \frac{V_2^2}{2g} + z_2 \qquad \text{(EQ 1.12)}$$

where p_1 and p_2 are pressures at the two points of observation, z_1 and z_2 are elevations of the two points of observation, H_a is the heat added to the system, H_l is the heat lost from the system during flow, and H_e is the heat extracted from the system.

Statistical relationships: Statistics deals with scientific methods for collecting, organizing, summarizing, presenting, and analyzing data, thereby enabling an individual to draw valid conclusions and make reasonable decisions. When collecting data, it is often impractical to observe an entire group, or population, because of its size; instead, a small portion of the group, or sample, is observed. If the sample is representative of the population, important conclusions about the population can be inferred from analysis of the sample.

The arithmetic mean, or average, of a set of numbers is equal to the sum of all of the numbers divided by the count of the numbers. One reason for taking measurements is to assess the degree of variation. Consequently, the degree to which numerical data tend to spread about an average is important. In mining, there are six possible causes of variation: (1) operator, (2) material, (3) equipment, (4) method, (5) tooling and wear, and (6) operating environment. Quality-control improvements can only be made if you can measure and predict variation in processes.

To fully describe data, the variation of the data from the mean must be known. Variance is an important measure of this variability in data. Sample variance is the average of the squared distance (deviations) of the data from the sample mean. The formula for sample variance (s^2) is

$$s^2 = \frac{\sum (X - \bar{X})^2}{n - 1} \qquad \text{(EQ 1.13)}$$

where:
    X = measurement in the sample
    $\bar{X}$ = sample mean
    n = sample size

The formula for population variance (σ^2) is

$$\sigma^2 = \frac{\sum (X - \mu)^2}{N} \qquad \text{(EQ 1.14)}$$

where:

    X = measurement in the population
    μ = population mean
    N = population size

Standard deviation is the most common measure of dispersion. The formula for sample standard deviation(s) is

$$s = \sqrt{\frac{\sum (X - \bar{X})^2}{n - 1}} \qquad \text{(EQ 1.15)}$$

where:

    X = measurements in population
    $\bar{X}$ = sample mean
    n = sample size

The formula for population standard deviation (σ) is

$$\sigma = \sqrt{\frac{\sum (X - \mu)^2}{N}} \qquad \text{(EQ 1.16)}$$

where:

    X = measurements in population
    μ = population mean
    N = population size

REFERENCES

Adler, L., and H.E. Naumann. 1970. *Analyzing Excavation and Materials Handling Equipment.* Research Division Bull. 53. Blacksburg, VA: Virginia Polytechnic Institute and State University.

American Geological Institute. 1997. *Dictionary of Mining, Mineral, and Related Terms.* Alexandria, VA: American Geological Institute.

Dasher, J. 1981. Using the modern metric system. *Mining Engineering* 33:2:147–149.

Peele, R. 1944. *Mining Engineers' Handbook.* New York: John Wiley and Sons, Inc.

Richardson, T.L. 1997. *Total Quality Management.* Albany, NY: Delmar Publishers.

PROBLEMS

Problem 1.1

A shuttle car travels at an average velocity of 300 fpm (feet per minute) from a continuous miner to a conveyor belt where it will discharge its payload. If travel time is 75 sec (seconds), what is the distance between the loading point and the dump?

Problem 1.2

After rounding a curve, a mine locomotive accelerates from a velocity of 3 mph (miles per hour) to 6 mph in 50 sec. What is the rate of acceleration in feet per second per second?

Problem 1.3

If a haulage truck at a surface mine accelerates at a uniform rate of 0.5 mphps (miles per hour per second) from rest, how fast will it be going in 20 sec?

Problem 1.4

What is the average velocity for the haulage truck in Problem 1.3 during the time interval between its initial velocity at 20 sec, 10 mph, and its final velocity at 1 min, 30 mph?

Problem 1.5

How far does the truck travel in Problem 1.4?

Problem 1.6

What is the truck's acceleration rate at the end of the interval for the situation in Problem 1.5?

Problem 1.7

A 20-ton mine locomotive is traveling at 6 mph (8.8 fps). Determine the retarding force of the brakes required to stop it in 105 ft on a level track.

Problem 1.8

A 10-ton mine locomotive is traveling at 6 mph (8.8 fps) when the motorman sees a fall of rock on the track ahead. He applies the brakes, but still hits the rock, moving it 3 in. What is the average force at impact?

Problem 1.9

Calculate the work done by a mine pump that discharges 300 gal of water into a sump located 27 ft above its intake.

Problem 1.10

A sedimentary rock weighs 160 lb per cu ft. What is its specific gravity?

Problem 1.11

The rock strata over a coal seam have an average specific gravity of 2.55. What is the pressure, in pounds per square inch (psi), that is exerted onto the seam if the depth of overburden is 400 ft?

Problem 1.12

A mine hoist lifts a cage weighing 11,000 lb that contains 11,500 lb of ore. The hoisting distance is 1,050 ft, and it takes 55 sec to travel this distance. The power supplied to the motor is 1,250 hp. Determine the work output, the power output, and power input in foot-pounds per second and the efficiency of the motor and hoist.

Problem 1.13

The following figure depicts the amount of coal that is left after mining (pillars) and the amount of coal that is extracted (rooms).

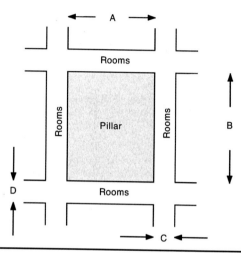

Illustrates Problem 1.13

A pillar's tributary area is defined as the sum of the pillar area and the mined-out area for one half of the distance to the adjacent pillars. If A = 60 ft, B = 75 ft, C = 20 ft, and D = 16 ft, determine the recovery ratio in a 6-ft coal seam.

Problem 1.14

Subsidence due to underground mining can cause considerable damage to surface structures. In Great Britain, for example, classification of structural damage depends primarily on changes in the length of structures.

If a 100-ft-long building is subjected to an increase in length of 0.2 ft (considered to be very severe damage) due to subsidence, what is the corresponding strain?

Problem 1.15

A surveyor's 100-ft-long steel tape has a cross section of 0.250 in. × 0.03 in. Determine its modulus of elasticity if its elongation is 0.064 in. when held taut by a force of 12 lb.

Problem 1.16

If air flows through an 18-in.-diam (diameter) ventilation duct at a velocity of 3,500 fpm, what is the quantity in cubic feet per minute?

Problem 1.17

Bituminous coal has a swell factor of 0.74. What is the volume of blasted coal, ready for loading, if the in-place volume is 1,200 cu ft?

Problem 1.18

A 50-ton mine locomotive typically has an adhesion factor of 0.20 when braking on clean, dry rails. Under these circumstances, what is its tractive effort in pounds?

Problem 1.19

What is the drawbar pull of the locomotive in Problem 1.18 if the locomotive resistance is 30 lb per ton?

Problem 1.20

What voltage would be required to force 5 amp (amperes) of current through a 40-ohm resistor in a d-c circuit?

Problem 1.21

Refer to the following diagram. If 75,000 cfm of air flows in split 1 and 25,000 cfm of air flows in split 2, what is the quantity of air flowing in split 3?

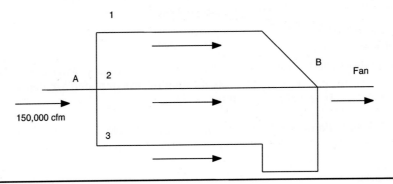

Illustrates Problem 1.21

Problem 1.22

What is the current flowing in the d-c circuit shown in the following diagram?

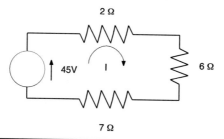

Illustrates Problem 1.22

Problem 1.23

A mass of air occupies 20 cu ft at 60°F and 14.7 psi. Determine its volume at 100°F and 12.68 psi.

Problem 1.24

A horizontal ventilation duct reduces in cross-sectional area from 1.40 sq ft to 0.80 sq ft. Assuming no losses, what pressure change will occur when 3.75 lb per sec of air flows? Assume w equals 0.0750 lb per cu ft for the pressure and temperature conditions involved.

Problem 1.25

Determine the arithmetic mean and the standard deviation for the following set (population) of numbers: 6, 3, 10, 5, 12, 7, 15, and 18.

PROBLEM SOLUTIONS

Solution 1.1

$s = v_{avg} t$
= (300 fpm)(75 sec)(1 min per 60 sec)
= 375 ft

Solution 1.2

3 mph = (3 mph)(5,280 ft per 1 mile)(1 hr per 3,600 sec) = 4.4 fps
6 mph = 8.8 fps

$a = (v_f - v_i)/t = [(8.8 - 4.4) \text{ fps}] / 50 \text{ sec} = 0.088$ ft per sec^2

Solution 1.3

$$v_f = v_i + at$$
$$= 0 + 0.5 \text{ mphps } (20 \text{ sec})$$
$$= 10 \text{ mph}$$

Solution 1.4

$$v_{avg} = (0.5)(v_i + v_f)$$
$$= (0.5)(10 + 30) = 20 \text{ mph}$$

Solution 1.5

10 mph = 14.67 fps (feet per second)
30 mph = 44.00 fps
$$s = (0.5)((14.67 + 44.00) \text{ fps})(40 \text{ sec})$$
$$= 1{,}174 \text{ ft}$$

Solution 1.6

$$s = v_i t + 0.5at^2$$
$$1{,}174 \text{ ft} = [(14.67 \text{ fps})(40 \text{ sec})] + [0.5 \, a \, (40 \text{ sec})^2]$$
$$1{,}174 \text{ ft} = 586.8 \text{ ft} + (0.5) \, a \, (1{,}600 \text{ sec}^2)$$
$$a = (587.2 \text{ ft})/(800 \text{ sec}^2)$$
$$a = 0.734 \text{ fpsps}$$

Solution 1.7

$$v_f^2 = v_i^2 + 2as$$
$$(0 \text{ fps})^2 = (8.8 \text{ fps})^2 + 2a(105 \text{ ft})$$
$$a = -0.369 \text{ fpsps}$$

$$f = \left(\frac{W}{g}\right)(a) = \left\{\frac{(20)(2{,}000)}{32.2}\right\}(0.369) = 458 \text{ lb}$$

Solution 1.8

Work = (force)(displacement)
Work = change in kinetic energy when velocity is changed abruptly

Thus,

$$(\text{force})(\text{displacement}) = \frac{1}{2}mv_1^2 - \frac{1}{2}mv_2^2$$

PRINCIPLES OF MINE PLANNING | 15

$$(\text{force})(0.25\text{ ft}) = \frac{1}{2}\left\{\frac{(10)(2{,}000)}{32.2}\right\}(8.8^2) - \frac{1}{2}\left\{\frac{(10)(2{,}000)}{32.2}\right\}(0^2)$$

$$\text{force} = 96{,}200\text{ lb}$$

Solution 1.9

$$\text{Weight of water} = \left(\frac{300\text{ gal}}{7.48\text{ gal per cu ft}}\right)(62.4\text{ lb per cu ft})$$

$$= 2{,}503\text{ lb}$$

$$\text{Work} = (\text{force})(\text{displacement})$$

$$= (2{,}503\text{ lb})(27\text{ ft})$$

$$= 67{,}581\text{ ft-lb}$$

Solution 1.10

$$\text{Specific gravity} = \frac{\text{weight of substance}}{\text{weight of water}}$$

$$= \frac{160\text{ lb per cu ft}}{62.4\text{ lb per cu ft}}$$

$$= 2.56$$

Solution 1.11

One cubic foot of rock weighs $(2.55)(62.4) = 159.12$ lb. Thus, the pressure exerted on the base of a cubic foot of this rock is equal to 159.12 lb. Since there are 144 sq in. in a square foot, the force can also be represented as

$$\frac{159.12\text{ lb}}{144\text{ sq in.}} = 1.1\text{ psi}$$

Therefore, 1.1 psi is exerted per foot of overburden. Since the overburden is 400 ft in this example, the pressure exerted on the coal seam is

$$(1.1)(400) = 440\text{ psi}$$

Solution 1.12

Work output: $[(11{,}000 + 11{,}500)\text{ lb}](1{,}050\text{ ft}) = 23{,}625{,}000$ ft-lb

Power output: $\dfrac{\text{work output}}{\text{time}} = \dfrac{23{,}625{,}000\text{ ft-lb}}{55\text{ sec}} = 429{,}545$ ft-lb per sec

Power input: $(1{,}250\text{ hp})\dfrac{(33{,}000\text{ ft-lb per min})}{(60\text{ sec per min})(1\text{ hp})} = 687{,}500$ ft-lb per sec

$$\text{Efficiency} = \frac{\text{power output}}{\text{power input}}$$

$$= \frac{429{,}545\text{ ft-lb per sec}}{687{,}500\text{ ft-lb per sec}}$$

$$= 62\%$$

Solution 1.13

$$\text{Recovery ratio} = \frac{\text{volume of mineral mined}}{\text{total volume of mineral prior to mining}}$$

$$= \frac{\text{volume in the tributary area} - \text{volume in the pillar}}{\text{volume in the tributary area}}$$

$$= \frac{(60+20)(75+16)(6) - (60)(75)(6)}{(60+20)(75+16)(6)}$$

$$= \frac{43{,}680 - 27{,}000}{43{,}680}$$

$$= 0.38 = 38\%$$

Solution 1.14

$$\text{Longitudinal strain} = \frac{\Delta l}{l} = \frac{0.2 \text{ ft}}{100 \text{ ft}} = 0.002$$

Solution 1.15

Stress = force per unit area

$$= \frac{(12 \text{ lb})}{(0.250 \text{ in.})(0.03 \text{ in.})}$$

$$= 1{,}600 \text{ psi}$$

Strain = (change in length)/initial length

$$= \frac{(0.064)}{(100)(12)} = 5.333 \times 10^{-5}$$

Modulus of elasticity = $(1{,}600 \text{ psi})(5.333 \times 10^{-5})$

$$= 30 \times 10^6 \text{ psi}$$

Solution 1.16

Duct area = πr^2

$$= \pi(0.75 \text{ ft})^2 = 1.767 \text{ sq ft}$$

Quantity = (cross-sectional area)(velocity)

$$= (1.767 \text{ sq ft})(3{,}500 \text{ fpm})$$

$$= 6{,}185 \text{ cfm}$$

Solution 1.17

$$\text{Loose volume} = \frac{\text{bank volume}}{\text{swell factor}}$$

$$= \left(\frac{1{,}200 \text{ cu ft}}{0.74}\right) = 1{,}622 \text{ cu ft}$$

Solution 1.18

Tractive effort = (weight of locomotive)(adhesion)

$$TE = (50 \text{ tons})(2{,}000 \text{ lb per ton})(0.20)$$
$$= 20{,}000 \text{ lb}$$

Solution 1.19

Drawbar pull = TE − locomotive resistance

$$= 20{,}000 \text{ lb} - [(50 \text{ tons})(30 \text{ lb per ton})]$$
$$= 18{,}500 \text{ lb}$$

Solution 1.20

Voltage = (current)(resistance)

$$= (5)(40) = 200 \text{ V}$$

Solution 1.21

According to Kirchhoff's first law,

$$Q_T = Q_1 + Q_2 + Q_3$$
$$150{,}000 \text{ cfm} = 75{,}000 \text{ cfm} + 25{,}000 \text{ cfm} + Q_3$$

Therefore, $Q_3 = 50{,}000$ cfm

Solution 1.22

According to Kirchhoff's second law,

$$V - (IR)_1 - (IR)_2 - (IR)_3 = 0$$
$$45 - 2I - 6I - 7I = 0$$

Therefore,

$$I = 3 \text{ amp}$$

Solution 1.23

A mass of air occupies 20 cu ft at 60°F and 14.7 psi. Determine its volume at 100°F and 12.68 psi.

$$\frac{P_1 V_1}{T_1} = \frac{P_2 V_2}{T_2}$$

$$\frac{(20 \text{ cu ft})(14.7 \text{ psi})}{(60 + 460)°R} = \frac{(V_2)(12.68 \text{ psi})}{(100 + 460)°R}$$

Therefore,
$$V_2 = 24.97 \text{ cu ft}$$

Solution 1.24

$$Q = \frac{3.75 \text{ lb per sec}}{0.0750 \text{ lb per cu ft}} = 50 \text{ cfs}$$

$$V_1 = \frac{Q}{A_1} = \frac{50 \text{ cfs}}{1.40 \text{ sq ft}} = 35.71 \text{ fps}$$

$$V_2 = \frac{Q}{A_2} = \frac{50 \text{ cfs}}{0.80 \text{ sq ft}} = 62.5 \text{ fps}$$

Applying the Bernoulli theorem,

$$\left(\frac{p_1}{w} + \frac{(35.71)^2}{2g} + 0\right) - 0 = \left(\frac{p_2}{w} + \frac{(62.5)^2}{2g} + 0\right)$$

Rearranging,

$$\frac{p_1}{w} - \frac{p_2}{w} = 40.86 \text{ ft of air}$$

$$p_1 - p_2 = 40.86 \text{ ft } (0.0750 \text{ lb per cu ft})$$

$$= 3.0645 \text{ psf}$$

$$= 0.0213 \text{ psi change}$$

Solution 1.25

The arithmetic mean is

$$\mu = \frac{\Sigma X}{N} = \frac{6+3+10+5+12+7+15+18}{8} = \frac{76}{8} = 9.5$$

The standard deviation is

$$\sigma = \sqrt{\frac{\Sigma (X-\mu)^2}{N}}$$

$$= \sqrt{\frac{(6-9.5)^2 + (3-9.5)^2 + (10-9.5)^2 + (5-9.5)^2 + (12-9.5)^2 + (7-9.5)^2 + (15-9.5)^2 + (18-9.5)^2}{8}}$$

$$= \sqrt{23.75}$$

$$= 4.87$$

CHAPTER 2

Mine Preplanning

Before planning a mine in detail, a preliminary analysis is often conducted. This preplanning phase enables the mining engineer to quickly determine whether or not a particular mining property warrants further consideration. In fact, a properly conducted preplanning phase can come within 10% to 20% of the actual results, thereby saving the mining engineer valuable time before committing to a fine-tuned analysis.

Among the many questions to be answered during mine preplanning are:

- What will the operating conditions for the mine be like?
- How long will it take to complete the project?
- What health, safety, and environmental regulations will affect the project?

To answer these questions, reserve estimation, production and staff planning, and project scheduling are calculated. The impact of regulations can only be assessed by understanding the basic science underlying these considerations. The following sections discuss these various topics of mine preplanning in greater detail.

RESERVE ESTIMATION

Estimation of the various characteristics of a reserve, such as quantity, grade, and thickness, is an ongoing process that lasts throughout the life of a mining venture. The reserve-estimation phase of mine planning, however, is more closely associated with the preplanning phase. There are many different techniques used in reserve estimation of solid mineral deposits; variations among these techniques are introduced because of such considerations as ease of application or level of accuracy. In any event, each technique is developed in the following manner:

1. The mineral body is represented pictorially and then subdivided into blocks related to one or more samples of exploration data.
2. The area and volume of each block is determined.
3. The volumes are converted to representative tonnages (e.g., grades).
4. The results are tabulated.

This procedure is illustrated using the polygonal (area-of-influence) technique. This technique is used because of its ease of application, flexibility in dealing with different deposits, and general acceptance.

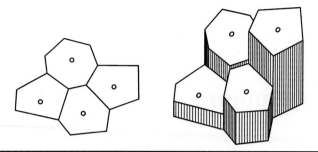

FIGURE 2.1 Method of polygons (plan and polygonal prisms)
Source: Popoff 1966.

TABLE 2.1 Computation of reserves by the polygonal method

Polygon Number	Area (S), sq ft	Thickness (t), ft	Volume (V), cu ft	Weight Factor (F), cu ft per ton	Raw Material Reserves (Q), tons	Valuable Component	
						Grade (c), %	Reserves (P), tons
1	S_1	t_1	V_1	F	Q_1	c_1	P_1
2	S_2	t_2	V_2	F	Q_2	c_2	P_2
3	S_3	t_3	V_3	F	Q_3	c_3	P_3
⋮	⋮	⋮	⋮	⋮	⋮	⋮	⋮
n	S_n	t_n	V_n	F	Q_n	c_n	P_n
Total	$\sum_{i=1}^{n} S$		$\sum_{i=1}^{n} V$		$\sum_{i=1}^{n} Q$		$\sum_{i=1}^{n} P$
Average		$t_{av} = \dfrac{\sum_{i=1}^{n} V}{\sum_{i=1}^{n} S}$				$c_{av} = \dfrac{\sum_{i=1}^{n} P}{\sum_{i=1}^{n} Q}$	

Source: Popoff 1966.

The polygonal technique assumes that all characteristics of a mineral body extend halfway between a point of observation and any other point of observation. The explored portion of the mineral body is represented by polygonal prisms whose depths relate to the characteristic in question (grade, thickness, etc.) and whose plane bases relate to the area of influence of each point of observation (drill hole, etc.). This is shown in Figure 2.1. The polygons must be constructed in a definite order, usually clockwise and from the periphery to the center of the deposit. Table 2.1 is then used to calculate the characteristics in question.

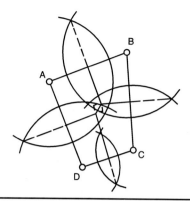

FIGURE 2.2 Construction of perpendicular bisectors
Source: Popoff 1966.

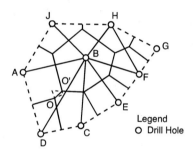

FIGURE 2.3 Correct construction of polygons by perpendicular bisectors
Source: Popoff 1966.

The polygonal method, by definition, assumes that the characteristics of a mineral body extend halfway between points of observation. Consequently, it becomes readily apparent that the evaluation of a vertical–drill-hole exploration program requires the sides of the polygons to be formed by the intersection of perpendicular bisectors (Figure 2.2). Figures 2.3 and 2.4 show correct and incorrect polygon construction. The criterion for correct construction is that all points within a particular polygon must be closer to the polygon's rallying point than those located in any other polygon. This is clearly not the case for certain points that are located in the polygon containing rallying point B, as shown in Figure 2.4.

Although this method can be used where drill holes are irregularly spaced, calculations become less tedious when a square-net (Figure 2.5) or chessboard (Figure 2.6) grid is adopted. The numbers shown in Figures 2.5 and 2.6 represent the relative weight for each of the shaded areas. In general, a greater number of polygons and a more continuous mineral body result in a more regular grid and a more accurate computation.

To illustrate the necessity to determine quality at the preplanning stage, consider how environmental regulations impact the marketing of coal. As a result of the 1990 Clean Air Amendments Act, only coal that emits less than 1.2 lb of SO_2 per million Btu may be burned by power plants, unless arrangements can be made for emissions allowances or

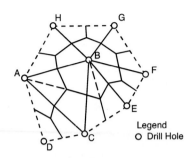

FIGURE 2.4 Incorrect construction of polygons
Source: Popoff 1966.

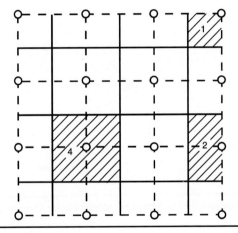

FIGURE 2.5 Square net pattern
Source: Popoff 1966.

the power plants use scrubbers (www.pollutionengineering.com). Coal that satisfies the requirement is designated as "compliance coal." To determine if a specific coal meets the requirement, the following equation is used:

$$\frac{\text{Percent sulfur} \times 20{,}000}{\text{Btu rating of the coal}} = \text{lbs of } SO_2 \text{ produced per million Btu} \qquad \text{(EQ 2.1)}$$

Clearly, an effective exploration program is crucial to determining the marketability of a mineral reserve.

PRODUCTION PLANNING

Before the production potential of a mine can be properly evaluated, a thorough exploration program should be put in place. It seems obvious that the nature and properties of the deposit and adjacent strata, the presence of impurities, and other mining conditions should be adequately determined prior to production planning and equipment selection.

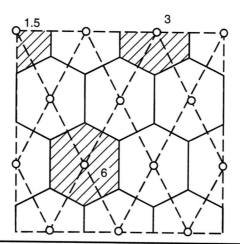

FIGURE 2.6 Chessboard pattern
Source: Popoff 1966.

A lack of sufficiently detailed information regarding many mining parameters has led to the failure of numerous operations. For example, targeted production tonnages and equipment sizes for a coal mine are intimately related to the seam thickness. Needless to say, the quality of information that is gathered during an exploration program can have a tremendous impact on the production plan.

Production plans are initially developed by determining reasonable shift tonnages per operating section. These values are usually developed using historical data (same company, ore body or seam; same company, similar ore body or seam) or from published information (another company, same ore body or seam; another company, similar ore body or seam). Industry journals and handbooks are also excellent sources for this kind of information. After determining an average section shift tonnage, a tentative mine plan can be developed to aid in calculating the rates of advance, the number of operating sections to be phased in until full production is reached, and the development time. For the latter, it is important to bear in mind that it may take several years to reach full production after development commences.

To avoid overestimation in production planning, daily working hours and annual working days should be selected conservatively. For example, an 8-hour shift can rapidly be reduced to less than 6 hours of actual face (mining) time when times for travel in/travel out, lunch, safety meetings, preparation to begin mining/preparation to end mining, and scheduled/unscheduled equipment downtimes are considered. When weekends, holidays, vacations, and unscheduled work stoppages are taken into account, an operating year can quickly be reduced to 220 to 250 days. In the past, underground mines were scheduled to operate 228 days per year (planning was simplified because this number is divisible by 12, i.e., 19 days per month). However, many modern underground mines have adopted work schedules with daily shifts exceeding 8 hours in length and 6- and 7-day weekly production schedules. Large surface mining operations are usually scheduled year-round at 24 hours per day.

STAFF PLANNING

The production planning phase of mine planning establishes the number of sections, daily operating shifts, annual operating days, and development time. Staff planning, on the other hand, is the phase of mine preplanning that assigns the number of hourly and wage personnel to the operation for each of the above-mentioned facets of production planning. As in production planning, the information necessary to develop staff requirements can often be obtained from historical and published sources. Bear in mind that different mining systems and methods often have different staffing requirements. Also, the number of hourly mineworkers who work in support of the operating sections (such as maintenance, haulage, and ventilation personnel) can be quite significant. In coal mines, for example, the ratio of outby (support) personnel to production personnel can be approximately 1:2; thus, a six-section mine that operates two shifts per day (12 continuous-miner shifts per day) and has six hourly employees per section may have as many as 220 mineworkers on the payroll.

Table 2.2 shows a typical staffing table for a room-and-pillar mine and a longwall coal mine, each with a specified annual tonnage. The longwall mine is based on one longwall unit, three continuous miners units driving the longwall gate roads, and one continuous miner unit driving the mains.

TABLE 2.2 Typical staffing for underground coal room-and-pillar and longwall mines

	Room-and-Pillar	Longwall
Longwall Crew (one unit):		
Shearer operator		6
Shield operator		5
Headgate operator		2
Utility		3
Belt attendant		3
Set-up		6
Subtotal:		25
Miscellaneous Underground:		
Continuous miner operator	8	12
Continuous miner helper	8	
Roof bolter	8	43
Roof bolter helper	8	
Shuttle car operator	24	25
Utility	8	12
Section mechanic	8	28
Supply motorman	10	6
Beltman	8	20
Wireman		2
Trackman	2	
Pumper	1	3

(table continues on next page)

TABLE 2.2 Typical staffing for underground coal room-and-pillar and longwall mines (continued)

	Room-and-Pillar	Longwall
General inside labor	12	100
Fire boss	4	9
Loader operator		12
Bratticeman		2
Timberman		3
Subtotal:	109	277
Miscellaneous Surface:		
Lampman	3	1
Front-end loader operator	5	2
Shop mechanic	3	
Utility	2	3
Labor, unskilled	3	
Dispatcher		3
Prep plant		40
Subtotal:	16	49
Salaried Personnel:		
Superintendent	1	1
Longwall coordinator		1
Assistant superintendent		1
General mine foreman	1	1
Assistant mine foreman		1
Section foreman	8	40
Additional supervisors		2
Maintenance superintendent	1	
Mine maintenance foreman		13
General shop foreman	1	
Chief mine engineer	1	1
Draftsman		1
Survey crew		2
Safety director	1	1
Safety inspector		2
Electrical engineer		1
Supply clerk		3
Shift foreman		5
Mine clerk		1
Warehouseman		2

(table continues on next page)

TABLE 2.2 Typical staffing for underground coal room-and-pillar and longwall mines (continued)

	Room-and-Pillar	Longwall
Chief belt foreman		1
Belt foreman		4
Master mechanic		1
Longwall maintenance coordinator		1
Electrical foreman		1
Shop foreman		1
Maintenance engineer		1
General plant foreman		1
Plant foreman		5
Timekeeper and bookkeeper	1	
Purchasing supervisor	1	
Subtotal:	16	95
Total Personnel:	141	446
Production Days per Year:	350	
Production per Day (Tons):	3,417	
Production per Year (Tons):	1,195,950	5,596,000
Productivity (Tons per Employee-year):	8,482	12,547

PROJECT SCHEDULING

When large-scale mining ventures are being planned, the mine-planning engineers' most pressing concern is to minimize project delays. The most widely used project scheduling technique is probably the critical path method (CPM). Because cost/time tradeoffs are also concerns in mine planning, a refinement of CPM, known as the program evaluation and review technique (PERT), is often used. When applying PERT to mine project scheduling, it is necessary to determine all of the significant activities in the project and the interrelationships among each activity. The completion of an individual project activity is referred to as an "event," and the relationships among all activities and events are referred to as the project's "network."

Figure 2.7 is an example of a project network, with the lettered nodes representing events and the arrows between events signifying activities. This network shows that there are four "paths" between the project initiation and the project completion events that must be followed: A–B–F–G, A–C–F–G, A–C–E–G, and A–C–D–G. The numbers between events (signifying days) in Figure 2.7 represent the "optimistic estimate," "most likely estimate (circled)," and "pessimistic estimate" for the completion of each activity. If only the most likely estimate is considered, it is readily apparent that path A–C–D–G takes the longest amount of time (24 days). This is referred to as the project's "critical path" because it is the longest path (in time) between the project's initiation and completion events. By defining the critical path, the mine planner is able to pinpoint areas where individual delays can hinder the overall project; specifically, any delay along the critical path will automatically postpone project completion. By the same token, the planner can determine if a delay in any other parallel path can be endured without hampering the

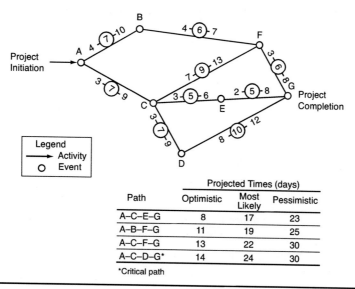

FIGURE 2.7 PERT program network
Source: Richards and Greenlaw 1966.

project's on-time completion date. In other words, a reallocation of staff or materials may reduce overall costs, thereby providing the cost/time trade-off mentioned earlier.

In summary, CPM and PERT are useful tools for scheduling large projects, such as the installation of a mining operation. Although computer programs may be necessary to handle large networks with many events and activities, the concepts described in this section are fundamental to even the most complicated project.

MINE SAFETY AND HEALTH

In addition to any relevant state and local regulations, mines in the United States are also subject to federal regulation; the US Department of Labor's Mine Safety and Health Administration (MSHA) is the regulatory agency. A list of federal regulations pertaining to mine safety and health are contained in the Code of Regulations (CFR). The specific volume is referred to as Title 30, which is further subdivided based by Part and Section for specific statutes. When Sections are grouped, they are described by a capital letter as a Subpart. Locating a specific regulation follows a shorthand description, which combines the Title with the appropriate numerical designation. For example, the location of the federal regulation (Title 30) that deals with mandatory health standards in underground coal mines (Part 70) with specific reference to respirable dust standards (Section 100) to be continuously maintained (a) is described as 30 CFR 70.100 (a). This convention is used in this text when it is necessary to call attention to a federal regulation.

To measure safety and health performance, MSHA uses several indices. Since 1978, mine operators have been required under 30 CFR 50 to submit reports of injuries, occupational illnesses, and related data to MSHA's Safety and Health Technology Center (SHTC) in Denver, Colorado. MSHA uses four principal indices for measuring mine safety: (1) fatal incidence rate (fatal IR), (2) nonfatal incidence rate resulting in days lost (NFDL IR), (3) injuries with no days lost incidence rate (NDL IR), and (4) severity measure (SM).

The incidence rate is defined as the number of injury occurrences per 200,000 miner-hours:

$$\text{IR} = \frac{(\text{number of injury occurrences})(200{,}000)}{\text{number of miner-hours}} \qquad \text{(EQ 2.2)}$$

which approximates the number of injury occurrences per year per 100 miners. A fatal injury results in death, an NFDL represents a nonfatal injury occurrence resulting in one or more days from the employee's scheduled work or days of restricted activity while at work, and an NDL represents a nonfatal injury occurrence resulting in medical treatment other than first aid. The severity measure represents the number of lost workdays per 200,000 miner-hours:

$$\text{SM} = \frac{(\text{number of lost workdays})(200{,}000)}{\text{number of miner-hours}} \qquad \text{(EQ 2.3)}$$

As an example, Tables 2.3 and 2.4 list the principal locations of the federal regulations addressing coal-mine safety and health that, in turn, affect coal-mine design.

Airborne Contaminants

According to 30 CFR 56.5001 and 30 CFR 57.5001, except for asbestos dust, the exposure to airborne contaminants in surface mines and underground metal and nonmetal mines shall not exceed, on the basis of a time-weighted average, the threshold limit values (TLVs) adopted by the American Conference of Governmental Industrial Hygienists (ACGIH).

For underground and surface coal mines, Subpart B of 30 CFR 70 and 30 CFR 71 indicates that each operator shall continuously maintain the average concentration of respirable dust in the mine atmosphere during each shift at or below 2.0 mg/m^3 of air. Further, when the respirable dust in the mine atmosphere of the active workings contains more than 5% quartz, the operator shall continuously maintain the average concentration of respirable dust in the mine atmosphere during each shift to which each miner in the active workings is exposed at or below a concentration of respirable dust, expressed in milligrams per cubic meter, computed by dividing the percent of quartz into the number 10.

Noise Control

Noise-induced hearing loss (NIHL) is an insidious condition because it usually begins as a gradual, progressive reduction in the quality of communication with other people and responsiveness to the environment and can ultimately result in permanent loss of hearing or deafness. This debility is often underrated, perhaps because there usually is no outwardly visible effect and, in most cases, little or no pain. This is significant in occupational settings, such as mining, because the inability to properly hear various environmental noises will result in diminished sensitivity to potentially hazardous sounds. The inability to discriminate between "sounds" that are meaningful and wanted and "noises" that are annoying and unwanted further diminishes occupational safety. Current concerns about NIHL among mineworkers are based on the fact that, even with decades of

TABLE 2.3 Subparts in 30 CFR that pertain to mandatory safety and health standards in underground coal mines

Mandatory Health Standards (30 CFR 70)
- A. General
- B. Dust Standards
- C. Sampling Procedures
- D. Respiratory Equipment
- E. Dust from Drill Rock
- F. Noise Standard

Mandatory Safety Standards (30 CFR 75)
- A. General
- B. Qualified and Certified Persons
- C. Roof Support
- D. Ventilation
- E. Combustible Materials and Rock Dusting
- F. Electrical Equipment - General
- G. Trailing Cables
- H. Grounding
- I. Underground High-Voltage Distribution
- J. Underground Low- and Medium-Voltage Alternating Current Circuits
- K. Trolley Wires and Trolley Feeder Wires
- L. Fire Protection
- M. Maps
- N. Blasting and Explosives
- O. Hoisting and Mantrips
- P. Emergency Shelters
- Q. Communications
- R. Miscellaneous
- S. Approved Books and Records

efforts to deal with the issue, the problem still persists. Much like other human senses, good hearing is often taken for granted until it is lost.

It should be noted that the human ear is not equally sensitive to all sound frequencies. To account for this, sound-level meters have frequency-response weighting networks that attenuate sounds of certain frequencies. This results in a weighted pressure level of total sound. Three scales are commonly used for these networks: A, B, and C. The A scale comes closest to approximating the ear's response characteristics; in other words, the A-weighting curve is an approximation of equal loudness perception characteristics of human hearing for pure tones relative to a reference of 40-dB sound pressure level at 1,000 Hz (Earshen 1986). The A scale also approximates the damage potential for high-level sounds. Regulatory agencies such as MSHA use this scale, and subsequent references to A-weighted sound levels will be in dBA.

When two or more sources of noise are in close proximity, their individual levels cannot be added directly, due to the logarithmic scaling of decibels. To determine the

TABLE 2.4 Subparts in 30 CFR that pertain to mandatory safety and health standards in surface coal mines and surface work areas of underground coal mines

Mandatory Health Standards (30 CFR 71)
- A. General
- B. Dust Standards
- C. Sampling Procedures
- D. Respirable Dust Control Plans
- E. Surface Bathing Facilities, Etc.
- F. Sanitary Toilet Facilities, Etc.
- G. Drinking Water
- H. Airborne Contaminants
- I. Noise Standard

Mandatory Safety Standards (30 CFR 77)
- A. General
- B. Qualified and Certified Persons
- C. Surface Installations
- D. Thermal Dryers
- E. Safeguards for Mechanical Equipment
- F. Electrical Equipment - General
- G. Trailing Cables
- H. Grounding
- I. Surface High-Voltage Distribution
- J. Low- and Medium-Voltage Alternating Current Circuits
- K. Ground Control
- L. Fire Protection
- M. Maps
- N. Blasting and Explosives
- O. Personnel Hoisting
- P. Auger Mining
- Q. Loading and Haulage
- R. Miscellaneous
- S. Trolley Wires and Trolley Feeder Wires
- T. Slope and Shaft Sinking
- U. Approved Books and Records

net effect of two noise sources, Table 2.5 is used. To use Table 2.5, take the dB readings of the two sources, find the difference between the two (column 1), and add the corresponding value in column 2 to the higher of the two readings. The result is the net effect of the two sources. When there are three or more sources, start with the two highest readings, find the net effect of that pair, and then use that calculated value of the pair with the next highest reading. Continue in this manner until all levels have been added.

MSHA regulates noise in metal and nonmetal mines (30 CFR 56.5050; 30 CFR 57.5050) and in coal mines (30 CFR 70.500 to 30 CFR 70.511; 30 CFR 71.800 to 30 CFR 71.805). Regardless of the mining method, MSHA establishes the permissible noise exposures, as shown in Table 2.6. In instances where the daily exposure is composed of

TABLE 2.5　Values for combining decibel levels

Difference Between Two Sound Levels (dB)	Amount to be Added to the Higher Sound Level (dB)
0–1	3
2–4	2
5–9	1
≥10	0

TABLE 2.6　Permissible noise exposures

Hours of Exposure	Sound Level for Slow Response (dBA)
8.0	90
6.0	92
4.0	95
3.0	97
2.0	100
1.5	102
1.0	105
0.5	110
0.25 or less	115

two or more periods of noise exposure at different levels, their combined effect shall be considered rather than the individual effect of each using the following equation:

$$\frac{C_1}{T_1} + \frac{C_2}{T_2} + \ldots + \frac{C_n}{T_n} \qquad \text{(EQ 2.4)}$$

C_n indicates the total time of exposure at a specific noise level and T_n indicates the total time of exposure permitted at that level, as indicated in Table 2.6. If the value calculated in Eq. 2.4 exceeds 1.0, the mixed exposure shall be considered to exceed the permissible exposure. Interpolation between the values shown in Table 2.6 may be determined using the following equation:

$$\log T = 6.322 - 0.0602\ SL \qquad \text{(EQ 2.5)}$$

where T is the time in hours and SL is the sound level in dBA.

QUALITY CONTROL

Control Charts

There is an ever-increasing emphasis on product quality in the mining industry. As a result, managers can use control charts to determine when a process has changed sufficiently to require steps be taken to rectify the situation. Problems such as deteriorating product grade, equipment component tolerance, or accident statistics occasionally arise whereby a manager must decide, often quickly, whether the problems are simply a result

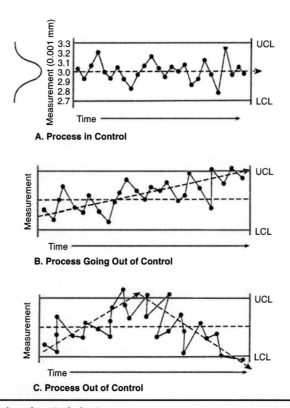

FIGURE 2.8 Examples of control charts
Source: Total Quality Management, Richardson, 1997. Reprinted with permission of Delmar Learning, a division of Thomson Learning.

of chance fluctuations or actual undesired changes in the process. Control charts provide a visual indicator to aid in decision making.

Control charts can be used for variables (means) or attributes (the number of defects). Figure 2.8 provides examples of control charts. Their construction frequently consist of a central line, target, or mean; an upper control limit (UCL) equal to the mean plus some multiple (usually 3 for 99.73% confidence) of the estimated standard error; and a lower control limit (LCL) equal to the mean minus the same multiple of the standard error (Figure 2.9). Points between these limits indicate normal or expected variation. A control chart is a run chart with statistically determined upper and lower limits drawn on either side of the process average. For example, Figure 2.8A reflects a process in control, Figure 2.8B reflects a process going out of control, and Figure 2.8C reflects a process out of control.

To develop a control chart for means, first determine the sample mean ($\bar{X}$). To have 99.73% confidence, you must be able to say that the sample mean lies in the range of $\mu - (3\sigma / \sqrt{N})$, which represents the LCL, to $\mu + (3\sigma / \sqrt{N})$, which represents the UCL, where μ is the population mean, σ is the population standard deviation, and N is the number of

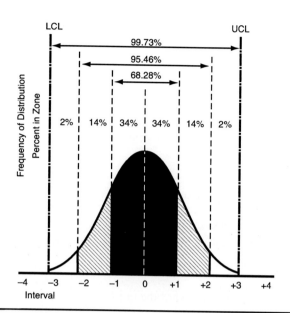

FIGURE 2.9 Normal distribution showing range of standard deviation
Source: Total Quality Management, Richardson, 1997. Reprinted with permission of Delmar Learning, a division of Thomson Learning.

samples taken at any one time. After the chart is constructed, plotting the sample means over time can provide a visual indication as to whether or not the system is in control.

To develop a control chart for the number of defects, first determine the sample mean ($\bar{X}$), which represents the average number of defects for each individual sampling. Next, calculate the proportion of the total samples represented by the defects ($\bar{p}$). To have 99.73% confidence, the LCL becomes

$$\bar{X} - [(3)(\bar{X} * (1 - \bar{p}))^{0.5}] \text{ , while the UCL becomes } \bar{X} + [(3)(\bar{X} * (1 - \bar{p}))^{0.5}].$$

REFERENCES

Baxter, C.H., and R.D. Parks. 1957. *Examination and Valuation of Mineral Property*. Reading, MA: Addison-Wesley Publishing Co., Inc.

Earshen, J.J. 1986. Sound Measurement: Instrumentation and Noise Descriptors. In *Noise & Hearing Conservation Manual*. Edited by E.H. Berger, W.D. Ward, J.C. Morrill, and L.H. Royster. Fairfax, VA: American Industrial Hygiene Association.

Popoff, C.C. 1966. *Computing Reserves of Mineral Deposits: Principles and Conventional Methods*. IC 8283. Washington, DC: US Bureau of Mines.

Richards, M.D., and P.S. Greenlaw. 1966. *Management Decision Making*. Homewood, IL: Richard D. Irwin, Inc.

Richardson, T.L. 1997. *Total Quality Management*. Albany, NY: Delmar Publishers.

www.pollutionengineering.com/archives/2002/0402/0402_Glossary.asp, 2002.

PROBLEMS

Problem 2.1

A mineral deposit is to be analyzed for mining based on the following information: (1) selling price of the refined metal: $0.75 per lb; (2) average ore grade: 0.5%; (3) recovery of metal from the ore after processing, smelting, and refining: 85%; (4) mining, milling, smelting, and refining costs expressed in terms of dollars per ton of ore: $6.50. Can the property be considered profitable at this time?

Problem 2.2

Given the following information on a 100-acre coal property, calculate its value: (1) seam thickness: 5 ft; (2) coal density: 80 lb per cu ft; (3) estimated mining recovery: 80%; (4) estimated yield of clean coal: 75% of run-of-mine; (5) Btu of clean coal: 12,000 Btu per lb; (6) selling price to the utility: $1.50 per million Btu.

Problem 2.3

A nine-entry main is to be driven toward the north end of a coal deposit (0.04 tons per cu ft) by a full-face continuous miner. The seam is 6 ft thick, the entries will be 15.5 ft wide, and the entries and breakthroughs will be driven on 100-ft centers on 90° angles. Production is estimated to be 1,000 tons per machine shift, and the mine operates two shifts per day, five days per week. A mining section (1 Left off of Main North) is to be prepared when the mains are extended 2,300 ft. Estimates indicate that as soon as the mains are driven past the proposed area for 1 Left, it will take 20 days to prepare the section (install overcasts, belt, etc.). Determine the most optimistic starting time, from the initiation of mains development, for the mining of 1 Left.

Problem 2.4

A coal deposit is to be developed using longwall mining. Each longwall panel will be 800 ft wide and 10,000 ft long. The seam thickness is 72 in., and the coal has a density of 80 lb per cu ft. The preparation plant's reject rate will be 20%. The mine has a contract requiring 15,000 tons of clean coal per day. A double-drum shearer will be used to extract the coal. With each pass, it will extract a 36-in. web of coal and will operate bidirectionally. The cutting speed in this coal is expected to average 40 linear fpm. A 2-min turn time is encountered after each pass. It is assumed that the shearer will actually be cutting coal for only 8 hr of the 10-hr shifts, due to scheduled and unscheduled downtimes. Annual major downtime is estimated to be 5 days per operating year. It will take 7 working days to move the longwall equipment from one face to the next. The mine will work two 10-hr shifts per day. Three continuous miner sections, which average 1,000 tons per day (each), will be used for development. The mine is scheduled to work 350 days per year.

Determine the daily tonnage required to meet the contract, the time required to mine one panel (excluding move time), and the number of longwalls required to be operational in this mine. Disregard any additional continuous-mining development for additional longwall panels.

Problem 2.5

An open-pit copper mine uses one 25-cu yd loading shovel to load 170-ton capacity haulage trucks. Usually, it takes seven passes with the shovel (cycle time: 40 sec per pass) to load each truck. The average load per truck is 80 yd. The copper ore density is 4,600 lb per cu yd (bank), and the ore has a swell factor of 0.67. The following times have been projected: haul time, 18 min; spot time, 0.5 min; dump time, 1 min; return time, 12 min. Calculate the production in tons per hour for each truck.

Problem 2.6

Using the method of polygons, determine the average seam thickness and tonnage represented by the following information if coal weighs 0.04 tons per cu ft:

Property Coordinates

East	North
1000	1000
7000	1000
1000	7000
7000	7000

Drill Hole	Seam Thickness, ft	East Coordinate	North Coordinate
1	4.3	2000	1000
2	6.2	4000	1000
3	5.8	6000	1000
4	5.7	1000	3000
5	5.3	3000	3000
6	6.1	5000	3000
7	4.9	7000	3000
8	4.8	2000	5000
9	5.9	4000	5000
10	5.4	6000	5000
11	6.0	1000	7000
12	5.3	3000	7000
13	5.5	5000	7000
14	5.7	7000	7000

Problem 2.7

Using the method of polygons, determine the average ore grade and tonnage represented by the following information if the cutoff grade is 1.8% and the ore weight is 1.7 tons per cu yd:

Drill Hole ID Number:	1	2	3	4	5	6	7	8	9	10	11	12	13	14	15	16
Collar elevation (ft):	1,000	1,050	1,060	1,010	1,020	1,070	1,050	1,000	1,040	1,020	1,000	1,070	1,040	1,010	1,020	1,000
East coordinate:	1,000	1,500	2,000	2,500	1,000	1,500	2,000	2,500	1,000	1,500	2,000	2,500	1,000	1,500	2,000	2,500
North coordinate:	1,000	1,000	1,000	1,000	1,500	1,500	1,500	1,500	2,000	2,000	2,000	2,000	2,500	2,500	2,500	2,500
Depth to mineralized zone:	100	150	160	110	120	170	150	100	140	120	100	170	140	110	120	100

Incremental Sequence of Samples Taken from the Mineralized Zone	Sample Grades in Percent															
10 ft	3.0	3.5	4.0	3.2	4.0	3.2	4.0	3.5	4.0	3.7	3.2	3.8	3.8	4.3	4.0	3.6
15 ft	2.0	3.0	4.2	2.7	3.5	2.8	3.5	3.0	3.6	3.8	3.0	3.9	3.2	4.0	3.8	3.0
20 ft	2.5	2.5	4.4	2.5	2.0	3.0	2.2	2.5	3.2	3.5	2.8	4.2	3.0	4.1	3.7	3.1
10 ft	3.5	2.0	4.5	3.2	2.2	3.0	2.0	2.0	3.4	3.4	2.6	4.1	3.4	3.4	3.2	3.1
5 ft	3.8	2.5	4.0	3.0	2.0	2.5	2.0	2.5	3.1	2.7	2.4	3.4	2.4	2.7	2.6	2.2
10 ft	3.0	3.0	3.5	2.5	1.7	2.0	1.6	3.0	2.6	2.3	2.0	2.7	1.7	2.4	1.6	2.0
5 ft	1.7	2.0	2.0	1.5	1.6	2.2	1.5	2.0	1.6	1.7	1.9	1.5	1.5	2.0	1.4	1.7
20 ft	1.6	1.5	1.6	1.4	1.5	1.4	1.2	1.5	1.2	1.4	1.5	1.2	1.5	1.2	1.1	1.4

North boundary of the property: 2500N.
South boundary of the property: 1000N.
East boundary of the property: 2500E.
West boundary of the property: 1000E.

Problem 2.8

An exploration program was conducted for a mineral property. The drill holes were irregularly spaced, with the following representing the property and drill hole coordinates:

Property Coordinates

East	North
1000	1000
2630	1000
1000	2630
2630	2630

Drill Hole	East Coordinate	North Coordinate
1	1110	2480
2	1370	2540
3	1860	2480
4	2310	2420
5	1490	2300
6	2580	2280
7	1880	2110
8	2350	2000
9	1050	2000
10	1420	1820
11	1980	1730
12	2430	1630
13	1160	1410
14	1730	1410
15	2150	1350
16	1440	1140
17	1940	1140
18	2540	1140

Plot the property and drill hole coordinates and construct the polygons that represent the areas of influence for each point of observation. Comment on your findings.

Problem 2.9

The network shown is to be used in planning the installation of an underground crusher station. Determine the total project time and the critical path.

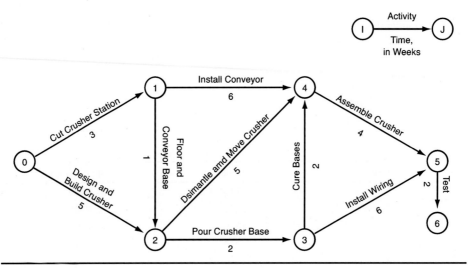

Illustrates Problem 2.9

Problem 2.10

A 10,000-tpd underground coal mine is in the planning stages. It is anticipated that salaried and supervisory personnel will constitute 20% of the work force. Each production crew will consist of eight miners and one foreman.

Several of the other mines operated by the company have been averaging 10 tons per miner per day and 200 tons per unit shift. However, there is reason to believe that this new mine will average 15 tons per miner per day and 300 tons per unit shift.

Determine the size of the required labor force for both rates of productivity.

Problem 2.11

Three coal seams are being compared. Seam A is mined in West Virginia and is rated at 0.75% sulfur and 12,200 Btu; seam B is mined in Illinois and is rated at 2.75% sulfur and 11,750 Btu; and seam C is mined in Wyoming and is rated at 0.2% sulfur and 8,750 Btu. Which of the three seams can be classified as a compliance coal?

Problem 2.12

A utility company is burning a bituminous coal rated at 12,000 Btu per lb. When burned, 38 lb of SO_2 is produced. Can this coal be characterized as a compliance coal?

Problem 2.13

During a recent year, three mines that use the longwall mining method recorded the following statistics:

	Mine A	Mine B	Mine C
Number of Fatalities	0	0	1
Number of NFDL Incidents	12	57	44
Average Number of Workers	359	486	619
Annual Employee-Hours	659,846	745,776	1,006,545
Annual Tonnage	6,226,681	2,020,303	3,517,821

Compare their fatal and nonfatal-with-days-lost incidence rates.

Problem 2.14

Refer to the previous problem: How many non-fatal lost workdays were recorded to provide a severity measure of 90 for mine A?

Problem 2.15

In 1996, there were 28 fatalities in US underground coal mines. The fatal severity measure was 371. If the total number of miner-hours for that year was 90,647,735, determine the standard lost workdays charged by MSHA for fatalities.

Problem 2.16

On a shop floor at a surface mine, there are three lathes in close proximity. Independently, lathe 1 emits a sound level of 74 dBA; lathe 2 emits a sound level of 81 dBA; and lathe 3 emits a sound level of 78 dBA. If all three lathes happen to operate at the same time, what would be the combined sound level, in dBA? What would be the combined sound level if lathe 1 is shut down?

Problem 2.17

During an 8-hour shift in a mineral processing plant, an employee is exposed to varying sound levels. Determine whether the allowable exposure is exceeded for the following sound levels and exposure times:

Sound Level (dBA)	Actual Exposure Time
85	2 hr
90	4 hr
95	1 hr
105	15 min
90	45 min

Problem 2.18

A mining equipment manufacturer is having a supplier provide ball bearings that have a mean diameter of 0.575 in. and a standard deviation of 0.008. In order to maintain quality control, the manufacturer is having the supplier sample six ball bearings every 2 hr of production, and the mean diameter is determined from each sample. The supplier operates 10 hours per day, so five samples of six ball bearings are evaluated on a daily basis. During one week, the mean sample diameters were as follows:

	9:00 a.m.	11:00 a.m.	1:00 p.m.	3:00 p.m.	5:00 p.m.
Monday	0.577	0.581	0.569	0.566	0.573
Tuesday	0.576	0.580	0.573	0.569	0.571
Wednesday	0.580	0.579	0.572	0.569	0.568
Thursday	0.572	0.567	0.563	0.565	0.563
Friday	0.567	0.565	0.568	0.563	0.565

Assuming a desired confidence level of 99.73%, develop a control chart and determine if the quality of the ball bearings is conforming to the requirements.

Problem 2.19

Assume that a component manufacturer, over a 4-day period, took six 500-component samples on a daily basis. The number of defective components for each sample is shown below:

	Day 1	Day 2	Day 3	Day 4
Sample 1	7	5	5	5
Sample 2	8	3	4	2
Sample 3	6	3	2	5
Sample 4	4	4	5	3
Sample 5	6	5	6	4
Sample 6	4	6	3	5

For a 99.73% confidence level, determine the average, the LCL, and the UCL for a control chart.

PROBLEM SOLUTIONS

Solution 2.1

1. Pounds of metal per ton of ore = (0.005)(2,000) = 10 lb.
2. Recovery after processing (10 lb)(0.85) = 8.5 lb.
3. Selling price of the refined metal recovered per ton of ore = (8.5 lb)($0.75) = $6.375.

The costs of mining and processing ($6.50) exceed the value of metal recovered ($6.375) on a per ton basis. Therefore, the property cannot be considered profitable at the present time.

Solution 2.2

Because 1 acre is equal to 43,560 sq ft, the in-place tonnage is as follows:

$$(100 \text{ acres})(43,560 \text{ sq ft})(5 \text{ ft}) \left[\frac{80 \text{ lb per cu ft}}{2,000 \text{ lb per ton}} \right] = 871,200 \text{ tons}.$$

The clean-coal recovery is as follows:

$$(871,200 \text{ tons})(0.8)(0.75) = 522,720 \text{ tons}.$$

Total Btu for the 522,720 tons is:

$$(522,720 \text{ tons})(2,000 \text{ lb per ton})(12,000 \text{ Btu per lb}) = 1.255 \times 10^{13} \text{ Btu}.$$

The property's value is, therefore:

$$(\$1.50/10^6 \text{ Btu})(1.255 \times 10^{13} \text{ Btu}) = \$18,817,920.$$

Solution 2.3

The layout for the mains is as shown below:

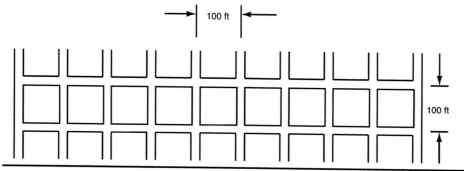

Illustrates Solution to Problem 2.3

To advance the mains for a distance of one breakthrough, the following tonnage must be extracted:

Mains: (9)(100 ft)(15.5 ft)(6 ft)(0.04) = 3,348
Breakthroughs: (8)(84.5 ft)(15.5 ft)(6 ft)(0.04) = 2,515
Total: 5,863 tons

To extend the mains for 23 breakthroughs (2,300 ft/100 ft) will require the following number of shifts: (23)(5,863 tons)/(1,000 tons) = 135 shifts. Thus, the total number of operating days is 68. When the 20 additional days for section preparation are added, a total of 88 days (approximately one third of one planning year) is required.

Solution 2.4

To produce 15,000 tons of clean coal per day, the mine must produce

(Clean tonnage ÷ recovery) = [15,000/(1.0 − 0.20)] = 18,750 raw tpd

One pass of the shearer:

Tonnage: (3 ft)(6 ft)(800 ft)(0.04 tons per cu ft) = 576 tons
Time: [(800 ft)/(40 fpm)] + 2.0 min = 22 min
Passes per shift: [(8 hr)(60 min)]/(22 min) = 21 passes
Tonnage per shift: (21 passes)(576 tons) = 12,096 tons
Panel advance per shift: (21 passes)(3 ft) = 63 ft
Time to extract one panel: (10,000 ft)/[(2) (63)] = 80 days
Total time per panel: 80 + [5(80/350)] + 7 = 89 days

Daily tonnage (including move and downtimes):

[(12,096 tons per shift)(80 days)(2 shifts per day)]/(89) = 21,745 tons
18,750 tpd − 3,000 tpd (from the continuous miners) = 15,750 tons
15,750 tpd ÷ 21,745 tpd = 0.72 longwall panels needed

Therefore, one panel should be sufficient to meet the tonnage required by the contract.

Solution 2.5

If the average load per truck is 80 cu yd, this is equivalent to the following tonnage:

(80)(4,600/2,000)(0.67) = 123 tons

Total cycle time:

Loading: [7(40)]/60 = 4.67 min
Spot time: 0.50 min
Haul time: 18.00 min
Dump time: 1.00 min
Return time: 12.00 min
Total: 36.17 min

[(123 tons per cycle)(60 min per hr)]/(36.17 min per cycle) = 204 tons per hr

Solution 2.6

Plotting the property and drill hole coordinates yields an exploration pattern identical to that shown in Figure 2.6. The area of influence for a full six-sided figure is calculated by subdividing the polygon into two triangles and one rectangle, as shown:

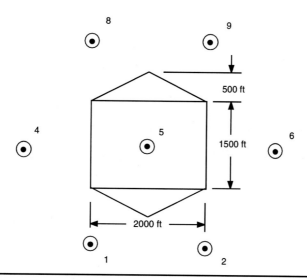

Illustrates Solution to Problem 2.6

Thus, the area of influence is:

$$2(0.5\ bh) + (bh) = 2[0.5((2,000)(500))] + [(1,500)(2,000)]$$

$$2\ triangles + rectangle = 4,000,000\ sq\ ft$$

This means that the area of influence for holes 5, 6, 8, 9, and 10 is 4,000,000 sq ft; the area of influence for holes 1, 2, 3, 4, 7, 12, and 13 is 2,000,000 sq ft; and the area of influence for holes 11 and 14 is 1,000,000 sq ft.

Tabulating the information:

Seam Thickness

Drill Hole	× (Area of Influence)	=	Volume
1	4.3 × 2,000,000	=	8,600,000
2	6.2 × 2,000,000	=	12,400,000
3	5.8 × 2,000,000	=	11,600,000
4	5.7 × 2,000,000	=	11,400,000
5	5.3 × 4,000,000	=	21,200,000
6	6.1 × 4,000,000	=	24,400,000
7	4.9 × 2,000,000	=	9,800,000
8	4.8 × 4,000,000	=	19,200,000

(table continues on next page)

44 | MINING ENGINEERING ANALYSIS

Seam Thickness (continued)

Drill Hole	x (Area of Influence)	=	Volume
9	5.9 × 4,000,000	=	23,600,000
10	5.4 × 4,000,000	=	21,600,000
11	6.0 × 1,000,000	=	6,000,000
12	5.3 × 2,000,000	=	10,600,000
13	5.5 × 2,000,000	=	11,000,000
14	5.7 × 1,000,000	=	5,700,000
	36,000,000 sq ft		197,100,000 cu ft

Average seam thickness = (197,100,000 cu ft)/36,000,000 sq ft = 5.48 ft

Tonnage = (197,100,000 cu ft)(0.04 tons per cu ft) = 7,884,000 tons

Solution 2.7

Plotting the property and drill hole coordinates yields an exploration pattern identical to that shown in Figure 2.5. Each hole has the following area of influence:

Drill Hole	Area of Influence (sq ft)
1	62,500
2	125,000
3	125,000
4	62,500
5	125,000
6	250,000
7	250,000
8	125,000
9	125,000
10	250,000
11	250,000
12	125,000
13	62,500
14	125,000
15	125,000
16	62,500

Drill Hole 1

Incremental Sequence	× Grade	= Weight
10	3.0	30
15	2.0	30
20	2.5	50
10	3.5	35
5	3.8	19
10	3.0	30
5	—	—
20	—	—
Average grade		194

The average grade for each drill hole is calculated in the following manner:

Average grade = (weight)/(ore depth) = 194/70 = 2.77

In a similar manner, the average grades for the other 15 holes are as follows:

Drill Hole	Weight	Ore Depth	Average Grade
2	202.5	75	2.70
3	301.0	75	4.01
4	194.5	70	2.78
5	164.5	60	2.74
6	207.5	75	2.77
7	166.5	60	2.78
8	202.5	75	2.70
9	233.5	70	3.34
10	234.5	70	3.35
11	200.5	75	2.67
12	265.5	70	3.79
13	192.0	60	3.20
14	266.5	75	3.55
15	216.0	60	3.60
16	205.0	70	2.93

The average ore grade and tonnage can then be determined with the following tabulation:

Average ore grade = (depth)(grade)/(ore depth) = 3,446.5/1,110 = 3.1%

Ore tonnage = (ore volume)(ore weight) = (5,810,184)(1.7) = 9,877,313 tons

MINING ENGINEERING ANALYSIS

Drill Hole	Average Grade	Ore Depth (ft)	Area of Influence (sq ft)	Ore Volume (cu yd)	Weight
1	2.77	70	62,500	162,037	194.0
2	2.70	75	125,000	347,222	202.5
3	4.01	75	125,000	347,222	301.0
4	2.78	70	62,500	162,037	194.5
5	2.74	60	125,000	277,778	164.5
6	2.77	75	250,000	694,444	207.5
7	2.78	60	250,000	555,556	166.5
8	2.70	75	125,000	347,222	202.5
9	3.34	70	125,000	324,074	233.5
10	3.35	70	250,000	648,148	234.5
11	2.67	75	250,000	694,444	200.5
12	3.79	70	125,000	324,074	265.5
13	3.20	60	62,500	138,889	192.0
14	3.55	75	125,000	347,222	266.5
15	3.60	60	125,000	277,778	216.0
16	2.93	70	62,500	162,037	205.0
		1,110		5,810,184	3,446.5

Solution 2.8

An accurate plot of the property and drill hole coordinates and construction of the polygons will yield an exploration map similar to the following:

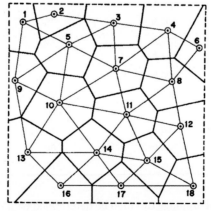

The persistence of each sample is assumed to extend halfway to all adjacent samples, and so the value of the property is calculated in a manner similar to the two previous problems. The area of each polygon could be measured either by a planimeter or by subdividing each polygon into triangles, finding their areas, and adding them together. Note, however, that the nonuniform approach to hole locations, particularly around the perimeter of the property, can lead to errors.

Solution 2.9

By inspection, it can be seen that there are 10 distinctly different activities that take the following durations:

Activity (i–j)	Duration (in weeks)
0 to 1	3
0 to 2	5
1 to 2	1
1 to 4	6
2 to 3	2
2 to 4	5
3 to 4	2
3 to 5	6
4 to 5	4
5 to 6	2

The network indicates that nodes 2, 4, and 5 are crucial because two or more paths feed into each. As such, it becomes necessary to determine the earliest and latest any activity can commence without holding up the entire project; this will expose the critical path. For example, there are two separate paths that terminate at node 2, namely 0-1-2 and 0-2. The path 0-1-2 is estimated to take 4 weeks to complete, while 0-2 is estimated to take 5 weeks. Thus, path 0-1-2 can start 1 week late (float one week) and still not hold up the completion date of node 2. Looking solely at these two paths, 0-2 is the critical path because its float is zero.

To determine the project's critical path and total time, the activity and duration listings previously shown should be expanded to include the earliest/latest start and the earliest/latest completion of each activity. This is shown in tabular form below.

Activity	Duration	Earliest Start (ES)	Earliest Finish (EF)	Latest Start (LS)	Latest Finish (LF)	Float (LF − EF)
0 to 1	3	0	3	1	4	1
0 to 2	5	0	5	0	5	0
1 to 2	1	3	4	4	5	1
1 to 4	6	3	9	4	10	1
2 to 3	2	5	7	6	8	1
2 to 4	5	5	10	5	10	0
3 to 4	2	7	9	8	10	1
3 to 5	6	7	13	8	14	1
4 to 5	4	10	14	10	14	0
5 to 6	2	14	16	14	16	0

The zero float activities are 0–2, 2–4, 4–5, and 5–6. This constitutes the critical path and, by looking at the earliest/latest finishing dates for node 6, it is apparent that the project time will be 16 weeks.

Solution 2.10

The total number of employees is determined in the following manner:

$$10 \text{ tons per man per day: } (10{,}000)/(10) = 1{,}000$$

$$15 \text{ tons per man per day: } (10{,}000)/(15) = 667$$

The number of salaried and supervisory personnel is 20% of the values determined in the preceding calculations:

$$10 \text{ tons per man per day: } 1{,}000(0.20) = 200$$

$$15 \text{ tons per man per day: } 667(0.20) = 134$$

The number of unit shifts is:

$$10 \text{ tons per man per day: } 10{,}000/200 = 50$$

$$15 \text{ tons per man per day: } 10{,}000/300 = 34$$

The number of unit shifts calculated required the following number of face miners and foremen:

$$10 \text{ tons per man per day: 50 foremen, 400 miners}$$

$$15 \text{ tons per man per day: 34 foremen, 272 miners}$$

Thus, the approximate labor allocation for the overall mine is as follows:

	10 Tons per Miner per Day	15 Tons per Miner per Day
Face miners	400	272
Foremen	50	34
Salaried	150	100
Backup miners	400	261
Total	1,000	667

Solution 2.11

Using Eq. 2.1, the three seams can be compared.

$$\frac{(\%S)(20{,}000)}{\text{coal Btu}} = \text{pounds of } SO_2 \text{ per } 10^6 \text{ Btu}$$

West Virginia: $[(0.75)(20{,}000)]/(12{,}200) = 1.23 \text{ lb}$
Illinois: $[(2.75)(20{,}000)]/(11{,}750) = 4.51 \text{ lb}$
Wyoming: $[(0.2)(20{,}000)]/(8{,}750) = 0.46 \text{ lb}$

Therefore, the West Virginia coal is just barely over the limit, and only the coal from Wyoming can be classified as a compliance coal.

Solution 2.12

As there are 2,000 lb in a ton, burning 1 ton of this coal will produce: (2000)(12,000) = 24 million Btu.

The total number of pounds of SO_2 produced in burning that same ton is 38 lb.

Because 38 lb of SO_2 are produced to provide 24,000,000 Btu, (38/24) = 1.58 lb of SO_2 are produced to provide 1,000,000 Btu. This coal cannot be classified as a compliance coal.

Solution 2.13

Using the following equation, the three mines can be compared:

$$IR = (\text{number of injury occurrences})(200{,}000)/(\text{number of miner-hours})$$

The fatal IRs for both mines A and B are 0.0. The fatal IR for mine C = (1)(200,000)/(1,006,545) = 0.2.

The NFDL IRs for the three mines are 3.64 for mine A, 15.29 for mine B, and 8.74 for mine C.

Solution 2.14

Using the following equation relating lost workdays and severity measure, the result is:

$$(\text{number of lost workdays}) = ((SM)(\text{number of miner-hours}))/(200{,}000)$$

$$= ((90)(659{,}846))/(200{,}000)$$

$$= 297 \text{ lost workdays}$$

Solution 2.15

Using the following equation, the severity measure can be computed:

$$SM = (\text{number of lost workdays})(200{,}000)/(\text{number of miner-hours})$$

Substituting the known values:

$$371 = (\text{number of lost workdays})(200{,}000)/(90{,}647{,}735)$$

Rearranging and rounding the answer to the nearest thousand:

$$\text{number of lost workdays} = (371)(90{,}647{,}735)/(200{,}000)$$

$$= 168{,}152 \text{ or } 168{,}000 \text{ workdays}$$

Because there were 28 fatalities, dividing that number into 168,000 results in 6,000 statutory days charged per fatality.

Solution 2.16

Looking at the 74-dBA and 78-dBA sources, there is a difference of 4.0. Referring to Table 2.5, adding 2.0 to 78 results in 80 dBA. Now, combining 80 with 81 requires the addition of 3.0 to 81. The final answer is 84 dBA.

When lathe 1 is shut down, the problem then becomes a combination of lathe 2 and lathe 3. Between the two, there is a difference of 3.0 dBA. Again, using Table 2.5, add 2.0 to the higher value. The final answer is 83 dBA.

Solution 2.17

Calculate the exposure versus permissible fraction using the values shown in Table 2.6 for each sound level and then add them together.

Sound Level (dBA)	Exposure Versus Permissible Fraction
85	0 (there is no limit for 85 dBA)
90	4/8 = 0.5 hr
95	1/4 = 0.25 hr
105	(15/60)/1 = 0.25 hr
90	(45/60)/8 = 0.09 hr

The sum of the exposure fractions results in 1.09. Because that value exceeds 1.0, the allowable exposure is exceeded.

Solution 2.18

If the mean diameter (μ) is 0.575, the control chart's LCL is $\mu - (3\sigma/\sqrt{N}) = 0.575 - (3(0.008)/2.449) = 0.565$, and the UCL is $\mu + (3\sigma/\sqrt{N}) = 0.575 + (3(0.008)/2.449) = 0.585$.

When the points representing the sample means are plotted, it appears that the process is in control through Wednesday. Starting on Thursday, however, the process appears to be going out of control, with three points beyond the LCL. An investigation of this trend is warranted.

Solution 2.19

Determine $\bar{X}$: Add all the defects and divide by the total number of samples taken: $110/[(6)(4)] = 4.6$.

Next, calculate $\bar{p}$: Since there were 110 defective components out of a possible 12,000 ($4 \times 6 \times 500$), the value for $\bar{p} = 110/12,000 = 0.009$.

The LCL becomes $\bar{X} - (3(\bar{X}(1-\bar{p})))0.5 = 4.6 - (3\sqrt{4.6 \times 0.991}) = -1.8$, which becomes 0.0. (Note: When the LCL is a negative number, the LCL becomes 0.0 when you construct the control chart.)

The UCL becomes $\bar{X} + (3(\bar{X}(1-\bar{p})))0.5 = 4.6 + (3\sqrt{4.6 \times 0.991}) = 11.0$.

CHAPTER 3

Ventilation

Traditionally, mine ventilation has been used to dilute, render harmless, and sweep away hazardous gases and dusts that may accumulate in the mine workings. In addition, the ventilation system must keep the mine air as close as possible to the composition of the outside air in order to provide a healthy environment for mineworkers and to prolong the useful life of mine equipment. When planning for a new mine, sound engineering dictates that both the maximum air quantity requirement and the acceptable mine pressure (head) loss related to the underground layout and projections be calculated.

Two problem areas that demand constant attention when calculating for efficient mine ventilation are: (1) the control of fugitive air (through short-circuiting or leakage) and (2) the provision of ample airway capacity. Fugitive air is the quantity of air that enters a mine but short-circuits the mine ventilation system through improperly constructed stoppings, doors, and overcasts. It represents an investment in power consumption (to move the air by the fan) while providing little, if any, return on investment. Typically, efficient mines are those that can deliver 50% or more of the air handled by their fans to the working faces. By paying close attention to a new mine's ventilation system at the outset, an engineer can provide numerous benefits during the later stages of mine development. It has often been noted that the primary cause of trouble in a mine's ventilation system can be found within a radius of 2,500 ft from the fan installation.

Ample airway capacity is essential to maintaining the mine's head loss within manageable limits. Similar to an industrial ventilation setting, the mine's headings are the ductwork for handling air. The difference between normal industrial ventilation and mine ventilation is apparent when more capacity is needed. A plant can install new or larger ductwork, but it is difficult for a mine to install new entries or enlarge existing entries. Air velocities are usually maintained below 800 fpm to limit the suspension of dust in the intakes. With this in mind, it is usually simple to estimate the number of entries required in the ventilation system. However, a problem often arises when the system's capacity is overestimated and the required head loss to achieve the desired quantity is excessive. Experience has shown that problems are often encountered when the mine's condition requires much more than 3 in. of water gage to ventilate the mine on a day-to-day basis. Thus, if proper attention is not paid to these parameters in the preplanning and initial development stages, it may take an extremely high water gage to achieve the ultimate quantity.

This chapter does not delve deeply into the theory of mine ventilation; many other texts have been written on that subject. Instead, the following sections point out the

FUNDAMENTALS OF AIRFLOW

Airflow Principles

The flow of air in a mine is induced by a pressure differential established between intake and exhaust shafts or openings. Ventilation pressures are usually described in inches of water gage, with 1 in. of water gage equal to a pressure of 5.2 lb per sq ft. Table 3.1 lists a number of additional principles of mine ventilation.

Pressure Losses

Pressure losses in mine ventilation systems are caused by friction between the heading walls and the air stream and by shock losses due to abrupt changes of the air stream caused by obstructions. The common pressure-loss formula for mine ventilation is:

$$H = \frac{KLOV^2}{(5.2)(A)} \qquad \text{(EQ 3.1)}$$

where H is the pressure loss in inches of water gage, K is the coefficient of friction, L is the length of the airway in feet, O is the airway perimeter in feet, V is the air velocity in feet per minute, and A is the cross-sectional area of the airway in square feet. Table 3.2 lists the US Bureau of Mines' (USBM) schedule of friction factors (K) for mine airways. These values were developed 75 years ago. Recent research indicates that the values of K are slightly less for coal mines than those presented in the USBM schedule, and Table 3.3 should be consulted in this instance. Note that all values shown in the tables must be multiplied by 10^{-10} to obtain the correct K values.

Recall from chapter 1 that the quantity (Q) is equal to the product of velocity and cross-sectional area. A substitution can be made into Eq. 3.1 to introduce quantity into the relationship:

TABLE 3.1 Airflow principles

- The pressure difference is caused by imposing some form of pressure at one point or a series of points in the ventilating system.
- The pressure created must be great enough to overcome frictional resistance and shock losses.
- Entries, both intakes and returns, serve as the air ducts.
- Airflow is directed from a point of higher pressure to a point of lower pressure.
- There is a square-law relationship between volumes and pressures; twice the volume requires four times the pressure.
- With respect to atmospheric pressure, mine ventilation systems are either exhausting (negative) or blowing (positive).
- The pressure drop for each split leaving from a common point and returning to a common point will be the same regardless of the air quantity flowing in each split.

Source: Kingery 1960.

VENTILATION | 53

TABLE 3.2 Bureau of Mines schedule for friction factors of mine airways

Values of $K \times 10^{-10}$

Type of Airway	Irregularities of Surfaces, Areas, and Alignment	Straight				Sinuous or Curved									
						Slightly				Moderately				High Degree	
		Clean (Basic Values)	Slightly Obstructed	Moderately Obstructed	Clean	Slightly Obstructed	Moderately Obstructed	Clean	Slightly Obstructed	Moderately Obstructed	Clean	Slightly Obstructed	Moderately Obstructed	Clean	Slightly Obstructed	Moderately Obstructed
Smooth-lined	Minimum	10	15	25	20	25	35	25	30	40	35	40	50			
	Average	15	20	30	25	30	40	30	35	45	40	45	55			
	Maximum	20	25	35	30	35	45	35	40	50	45	50	60			
Sedimentary (or coal)	Minimum	30	35	45	40	45	55	45	50	60	55	60	70			
	Average	55	60	70	65	70	80	70	75	85	80	85	95			
	Maximum	70	75	85	80	85	95	85	95	100	95	100	110			
Timbered (5-ft centers)	Minimum	80	85	95	90	95	105	95	100	110	105	110	120			
	Average	95	100	110	105	110	120	110	115	125	120	125	135			
	Maximum	105	110	120	115	120	130	120	125	135	130	135	145			
Igneous rock	Minimum	90	95	105	100	105	115	105	110	120	115	120	130			
	Average	145	150	160	155	160	165	160	165	175	170	175	195			
	Maximum	195	200	210	205	210	220	210	215	225	220	225	235			

Source: Kingery 1960.

MINING ENGINEERING ANALYSIS

TABLE 3.3 Friction factors (K) for coal mine airways and openings

	Value of K × 10⁻¹⁰					
	Straight			Curved		
Type of Airway	Clean	Slightly Obstructed	Moderately Obstructed	Clean	Slightly Obstructed	Moderately Obstructed
Smooth-lined	25	28	34	31	30	43
Unlined (rock bolted)	43	49	61	62	68	74
Timbered	67	75	82	85	87	90

Source: Ramani 1992.

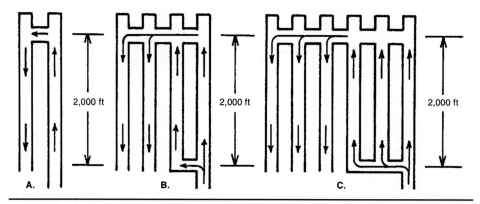

FIGURE 3.1 Single, double, and triple airways
Source: Adapted from Montgomery 1936.

$$H = \frac{KLO(Q^2/A^2)}{(5.2)A} = \frac{KLOQ^2}{(5.2)A^3} \quad \text{(EQ 3.2)}$$

The term $KLO/(5.2A^3)$ is often referred to as the resistance factor (R) of the airway. By substituting R into Eq. 3.2, the following standard formula can be derived:

$$H = RQ^2 \quad \text{(EQ 3.3)}$$

Parallel Flow in Airways

Suppose that a certain quantity (Q) of air has to be provided to a certain working area. Recalling that ample airway capacity is a major concern in mine ventilation, a study is to be conducted to determine the effects on the overall head loss in the intakes if one, two, or three intakes are provided (Figure 3.1).

In this instance, the values of K, L, 5.2, and Q in Eq. 3. 2 are unaffected by the number of entries, so all values will be combined into a constant, X. However, O and A will be affected in the following manner:

Substituting these values into Eq. 3.2 yields the following:

Number of Intakes

	1	2	3
O:	O	2O	3O
A:	A^3	$(2A)^3$	$(3A)^3$

1 intake: $(XO)/A^3$
2 intakes: $2(XO)/8A^3 = 1/4[(XO)/A^3]$
3 intakes: $3(XO)/27A^3 = 1/9[(XO)/A^3]$

The advantage of multiple parallel airways is readily apparent; the new resistance will be in the ratio of:

$$\left(\frac{1}{n_{new}/n_{original}}\right)^2 \qquad \text{(EQ 3.4)}$$

where n is the number of parallel airways.

Splitting of Air Currents

To capitalize on the benefit of having multiple parallel airways, it is often desirable to split the total quantity of air flowing through a mine into separate districts. The overall mine head can be reduced by splitting, because the pressure drop for each split leaving from a common point and returning to a common point will be the same regardless of the air quantity flowing in each split. To achieve the most benefit, the splits should be made as close as possible to the mine intake point and the several branches should reunite as close as possible to the mine exhaust point. In reference to Figure 3.2, it is assumed that Q_{total} is to be apportioned to three separate mining sections (splits). The air will split at a common location just inby the intake point and will recombine at a common location just inby the exhaust point. Because some splits are longer, wider, and

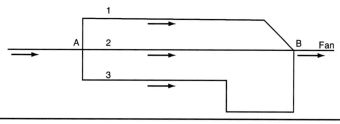

FIGURE 3.2 Splitting of airflow
Source: Kingery 1960

MINING ENGINEERING ANALYSIS

higher or have a different K value than the others, the air tends to seek the path of least resistance out of the mine. Rearranging Eq. 3.2 to determine the Q handled by each split and recalling that the H for each split is equal and can be cancelled from the formula, the following relationship for the potential air splitting can be developed:

$$q_{split} = \left[\frac{(5.2)(A_{split})^3}{(KLO)_{split}}\right]^{0.5} \quad \text{(EQ 3.5)}$$

Calculating q_{split} for each split, summing all of them, and multiplying each by the factor $[Q_{total}/(\text{summation of each individual } q_{split})]$ gives the quantity of air that will flow unregulated down each airway.

However, it is not normal practice to allow air to apportion itself in a mine ventilation system; usually a fixed amount of air must be delivered to each split. Using the example from the previous case, Q_{total} divides itself evenly among the three splits, and direct substitution into Eq. 3.2 establishes the head loss for each split. The split with the highest value of H is called the free split. This means that the other splits must be regulated by building a stopping with a sliding door perpendicular to the airways and just inby their common junction at the exhaust. This artificial barrier can be used to raise the head losses up to that found in the free split by sliding the doors so that the openings are reduced in size, thereby guaranteeing control over the apportionment of the total quantity. The necessary cross-sectional area of the opening in the regulator can be determined with the following equation:

$$A = \frac{40Q}{H^{0.5}} \quad \text{(EQ 3.6)}$$

where A is the cross-sectional area of the regulator's opening in square feet; Q is the air quantity, expressed in units of 100,000 cfm; and H is the pressure drop across the regulator in inches of water gage.

FAN SELECTION

Fans induce the airflow in mines and obey the fan laws listed in Table 3.4.

The performance of a fan in a mine ventilation system is a function of its characteristic curve, as established by the manufacturer. Because the mine operator controls the layout and maintenance of the airways and, thus, the mine characteristic, the operating point for the entire system becomes a function of fan design and mine design.

Table 3.4 can be used for calculations of fan performance when there are changes in the variables such as fan speed and air density. When two fans are compared, the variable dealing with their respective diameters enters into the analysis. For example, when

TABLE 3.4 Fan laws

Parameters	Speed Change (w,D are constant) $SR = n_2/n_1$	Diameter Change (n,w are constant) $D_iR = D_2/D_1$	Density Change (D,n are constant) $DR = w_2/w_1$
Quantity	SR	$(D_iR)2$	1
Head	$(SR)2$	1	DR
Power consumed	$(SR)3$	$(D_iR)2$	DR
Efficiency	1	1	1

Source: Kingery 1960.

w = air density in lb per cu ft; D = fan diameter in in.; n = fan speed in rpm; SR = speed ratio; D_iR = diameter ratio; DR = density ratio

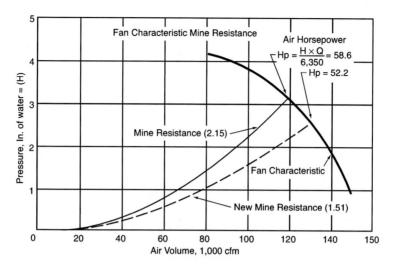

FIGURE 3.3 Relationship between fan characteristic and mine resistance
Source: Kingery 1960.

an 8-ft-diameter fan is compared to a 6-ft-diameter fan, the new (8-ft) fan's performance parameters will differ from those of the original (6-ft) fan by the following amounts:

$$D_iR = D_2/D_1 = (8)(12)/(6)(12) = 1.33$$

$$\text{Quantity}_{new} = Q_2 = (1.33)^2(Q_1)$$

$$\text{Head}_{new} = H_2 = (1.00)(H_1)$$

$$\text{Power consumed}_{new} = P_2 = (1.33)^2(P_1)$$

$$\text{Efficiency}_{new} = E_2 = (1.00)(E_1)$$

Q_1, H_1, P_1, and E_1 are the original values for quantity, head, power consumed, and efficiency, respectively.

Figure 3.3 shows the relationship between the mine and fan characteristic curves. Because a mining operation is dynamic, fan manufacturers provide several blade settings on their fans so that slight adjustments can be made to correspond to changes in mine resistance. Unfortunately, many operators use this flexibility as a quick way to overcome poor design and/or poor maintenance of roadways. Needless to say, fan adjustments should not be the first step taken to maintain acceptable quantities. As shown in Figure 3.3, driving additional airways can help to reduce mine resistance, lowering the required water gage and increasing air volume. The same results can be achieved if entries are driven straight, roof falls are cleaned up, and water is not permitted to gather in airways.

Although fans can be placed in series to increase pressure or in parallel to increase quantity, such changes are accompanied by an increase in power consumption. Again, proper planning is crucial to the success of a mine ventilation system.

VENTILATION REQUIREMENTS FOR A NEW MINE

Development of the ventilation layout for a modern mine is a complicated, time-consuming process that requires the use of a computer. However, certain aspects of preplanning can be conducted using the simple relationships presented in this chapter. Most new mines usually require a new source of intake and/or exhaust (i.e., shafts) for approximately every 2 miles of mine development, due to a gradual increase in mine resistance. Consequently, the following discussion applies only to the mine's requirements up to that point.

The first step in determining the ventilation requirements for a new mine is to calculate the maximum number of working sections. Once determined, the minimum quantity of air to be delivered to each working section must be established (in coal mines, this is approximately 10,000 cfm). Air will be lost in the various working sections because these areas are constantly changing and the stoppings installed are usually constructed of cinder blocks with nonmortared joints. Allow for a 50% loss of air delivered to the end of each panel due to leakage. If it is assumed that 50% of the air entering the mine will short-circuit the system, particularly within 2,500 ft of the mine entrances, doubling the amount delivered per panel should compensate for this loss. To restate this on a per panel basis, if 10,000 cfm makes it to the working faces, 20,000 cfm must be delivered to the panel. This, in turn, requires 40,000 cfm at the fan, which roughly translates into a 25% efficiency. Though conservative, it is probably better to underestimate, rather than to overestimate, the system's potential.

The next step is to calculate the required number of entries. If the maximum air velocity in the intakes is to be maintained at 800 fpm, the number of intakes required to handle the Q_{total} is:

$$N = (Q_{total} \text{ in cfm})/[(800 \text{ fpm})(A)] \qquad \textbf{(EQ 3.7)}$$

where A is the cross-sectional area of one airway in square feet. Note that N is always rounded up to the next highest integer. For symmetry, a similar number of return airways usually accompanies the intakes, although occasionally in large mines, one or two fewer returns may be driven as a cost-cutting measure with minimal detrimental consequences.

Finally, a line diagram of the intended mine plan is drawn (Figure 3.4). Once completed, the diagram should be analyzed to determine when and where the highest head loss is likely to occur. This is usually at the point where most mining sections are at the greatest distance from the air shafts. By estimating the air losses due to leakage in the mains, submains, and panels, the head loss for the free split and thus the entire mine can be approximated. A value that greatly exceeds 3 in. of water gage is a rapid indication of the shortcomings of the mine ventilation plan (i.e., too few entries, too great a distance between air shafts, etc.).

The solved problems in this chapter can be used as guides to aid in the analysis of mine ventilation layouts.

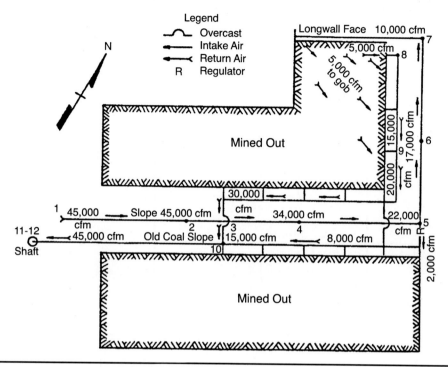

FIGURE 3.4 Typical line diagram for a mine ventilation system
Source: Kingery 1960.

REFERENCES

Hartman, H.L. 1982. *Mine Ventilation and Air Conditioning.* New York: John Wiley and Sons, Inc.
Kingery, D.S. 1960. *Introduction to Mine Ventilating Principles and Practices.* Bulletin 589. Washington, DC: US Bureau of Mines.
Montgomery, W.J. 1936. *Theory and Practice of Mine Ventilation.* Columbus, OH: The Jeffrey Manufacturing Co.
Ramani, R.V. 1992. Mine Ventilation. In *SME Mining Engineering Handbook.* Littleton, CO: Society for Mining, Metallurgy, and Exploration, Inc.

PROBLEMS

Problem 3.1

If the quantity of air flowing through a duct is 4,500 cfm and the manometer reading is 0.202 in. of water gage, what quantity corresponds to a manometer reading of 0.050 in.?

Problem 3.2

In the exhaust ventilation system shown in the following figure, determine all of the missing quantities for individual segments of airway.

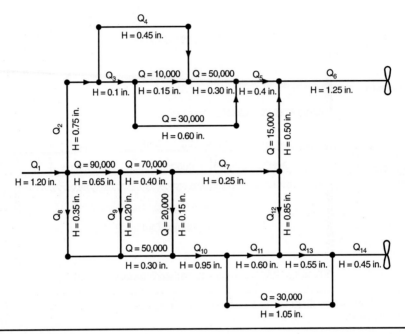

Illustrates Problem 3.2

Problem 3.3

Given the following mine ventilation network, determine the quantities flowing in the unknown segments and the "free split."

Segments	Q (cfm)	H_{loss} (in. water gage)
A→H	150,000	1.0
H→I	40,000	0.2
H→B	75,000	1.0
H→K	?	0.3
B→C	40,000	0.2
K→I	15,000	0.2
K→L	?	0.3
I→J	?	0.1
L→J	12,000	0.2
L→M	?	0.3
J→G	?	0.1

(table continues on next page)

Segments	Q (cfm)	H_{loss} (in. water gage)
B→D	20,000	0.2
D→C	?	0.1
B→F	?	2.0
C→E	?	0.2
D→E	10,000	0.3
F→E	?	1.0
F→G	10,000	0.2
E→M	?	3.0
G→M	?	0.4
M→N	?	1.0

Problem 3.4

In the ventilation system shown in problem 3.2, the static pressure drops between points in the network are indicated. Locate regulators where needed and calculate the amount of regulation (head loss in inches) needed. Determine the static head for each of the two fans.

Problem 3.5

Determine the size of the regulator to be placed across the heading containing Q_{12} in problem 3.4.

Problem 3.6

If the resistance factor for three entries is R, what will be the resistance factor for five entries?

Problem 3.7

A total flow of 180,000 cfm is allowed to distribute itself according to natural splitting among the following four parallel splits, each having K = 100×10^{-10}:

Split	Cross Section (ft)	Length (ft)
A	5 × 20	2,000
B	7 × 15	1,500
C	6 × 12	1,250
D	7 × 18	1,900

What will be the distribution of air?

Problem 3.8

A three-entry development section utilizing belt haulage is driven to a limit of 2,500 ft. If the entries are driven 18 ft wide in the 5-ft coal seam, what will be the pressure drop in the panel and the air horsepower if the air quantity is 40,000 cfm? Assume that the leakage is negligible.

K: intake: 60×10^{-10} return: 80×10^{-10}

62 | MINING ENGINEERING ANALYSIS

Problem 3.9

A panel entry is driven 5,000 ft deep employing diesel shuttle car haulage (no belts) in a three-entry, two-split configuration. Because of excessive gas liberation, the maximum amount of air that good mining practice will allow (to avoid the suspension and transportation of dust) will be delivered. Because of bad roof conditions, No. 1 (return) and No. 2 (intake) can be driven only 15 ft wide, whereas No. 3 entry (return) will be held to 12 ft wide in the 4-ft seam. The K values for entries No. 1 and No. 2 are found to be 60×10^{-10}, whereas that for No. 3 is 100×10^{-10}. Determine the flows as well as the panel air pressure requirement if the return splits are regulated in such a manner to produce equal flows in each return.

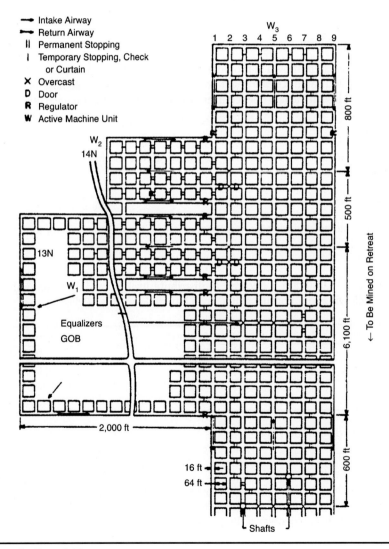

Illustrates Problem 3.10

Problem 3.10

Referring to the figure on the previous page, what will be the pressure drop in 14 North when it is driven to its full distance (2,000 ft) if the air quantity is 20,000 cfm at the last open crosscuts and no leakage is assumed? Remember that belt haulage is used in the panel, the seam thickness is 6 ft, and other pertinent dimensions are shown.

$$K: \text{intake: } 55 \times 10^{-10}$$

$$\text{return: } 75 \times 10^{-10}$$

Problem 3.11

Using the following ventilation schematic, determine how 75,000 cfm will apportion itself among the three splits. If 25,000 cfm is desired for each (neglect leakage), which will be the free split, how much regulation will be required, and where should the regulators be located?

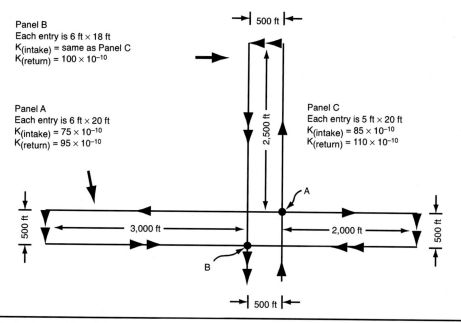

Illustrates Problem 3.11

Problem 3.12

A mine fan has the characteristic curve shown.

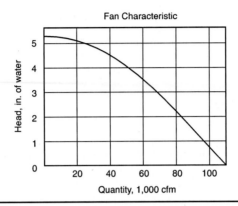

Illustrates Problem 3.12

The fan must supply 60,000 cfm at 2.5 in. static head. If the fan is currently operating at 800 rpm, what speed will be required to meet these conditions?

Problem 3.13

What would be the speed of the fan in problem 3.12 if it must supply air at 0.065 lb per cu ft? The standard density of air for mine ventilation work is considered to be 0.075 lb per cu ft.

Problem 3.14

A mine is presently operating at Q = 110,000 and H = 2.15 in. The fan (fan A) has the characteristic shown.

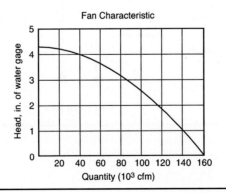

Illustrates Problem 3.14

Mine management is planning on developing three more sections. It was calculated that, at full development, the ventilation system would have to supply 250,000 cfm at approximately 2.50 in. A second available fan (fan B) was tested, and the following results were achieved:

Head (in.)	Quantity (cfm)
4.0	0
3.0	110,000
2.0	160,000
1.0	200,000
0.0	230,000

Is the company able to use this other fan?

Problem 3.15

The mine plan shown for problem 3.10 indicates that three of the six continuous-miner units of this medium-sized, underground coal mine are located in the following three sections: (1) Pillaring 13N, (2) Developing 14N, and (3) Developing Main North. The other three units are doing similar work at the southern end of the property.

The mine is located in a 6-ft thick seam where the depth of overburden is 1,000 ft. The intake air shaft is 17.5 ft in diameter, and the return air shaft is 18.0 ft in diameter. Each working section is composed of two splits of air with 20,000 cfm delivered to the last open crosscut of each split.

In planning for the future air requirements of the mine, it is anticipated that the worst case situation for the delivery of air to the mine will exist when the unit in 13N is transferred to 15N (thus becoming the free split). At the same time that 15N will be fully developed, the unit in 14N will be on retreat, and the unit in Main North will be transferred to develop the most inby panel on the right-hand side of Main North (16N).

Leakage of air will be apportioned as follows:
1. Twenty thousand cubic feet per minute will be lost in each section.
2. Thirty five thousand cubic feet per minute will be lost along Main North between the base of the intake air shaft and a point opposite the regulator of the bleeder system.
3. Forty thousand cubic feet per minute will be lost along Main North between the point which is opposite from the regulator of the bleeder system and 14N.
4. Fifteen thousand cubic feet per minute will be lost along Main North between 14N and 15N.
5. Of the air arriving at the last open crosscuts of 14N, it is assumed that 20,000 cfm will be directed into the bleeder system.

Determine the approximate overall quantity and head for this mine if the friction factors developed for slightly obstructed, straight, and unlined entries (Table 3.3) are assumed. Further, the friction factor for both air shafts should be assumed to be 15×10^{-10}.

66 | MINING ENGINEERING ANALYSIS

PROBLEM SOLUTIONS

Solution 3.1

$$H_1 = R_1 Q_1^2$$

$$H_2 = R_2 Q_2^2$$

because, $R_1 = R_2$ (the same duct),

$$\frac{H_1}{Q_1^2} = \frac{H_2}{Q_2^2}$$

Rearranging:

$$Q_2 = \left(\frac{H_2 Q_1^2}{H_1}\right)^{0.5}$$

$$= \left[\frac{(0.05)(4{,}500^2)}{(0.202)}\right]^{0.5}$$

$$= 2{,}239 \text{ cfm}$$

Solution 3.2

Recalling Kirchhoff's laws from chapter 1, the sum of the quantities entering a junction is equal to the sum of the quantities leaving the junction. Thus, the following quantities can be directly determined:

$Q_1 = 200{,}000$	$Q_8 = 30{,}000$
$Q_2 = 80{,}000$	$Q_9 = 20{,}000$
$Q_3 = 40{,}000$	$Q_{10} = 70{,}000$
$Q_4 = 40{,}000$	$Q_{11} = 40{,}000$
$Q_5 = 80{,}000$	$Q_{12} = 35{,}000$
$Q_6 = 95{,}000$	$Q_{13} = 75{,}000$
$Q_7 = 50{,}000$	$Q_{14} = 105{,}000$

Solution 3.3

To determine the unknown quantities, apply Kirchhoff's law, which states that the quantities flowing into a junction have to equal the quantities flowing from a junction. Therefore:

$$\text{Junction H: } (A \rightarrow H) = (H \rightarrow B) + (H \rightarrow I) + (H \rightarrow K)$$

$$150{,}000 = 75{,}000 + 40{,}000 + (H \rightarrow K)$$

Therefore, (H→K) = 35,000 cfm. If you proceed in a similar manner through the rest of the ventilation network, you should come up with the following values:

Segments	Q (cfm)
H→K	35,000
K→L	20,000
I→J	55,000
L→M	8,000
J→G	67,000
D→C	10,000
B→F	15,000
C→E	50,000
F→E	5,000
E→M	65,000
G→M	77,000
M→N	150,000

To determine the "free split" or path with the greatest head loss, indicate all of the various paths that air can take to get from A to N, add up each segment's head loss for each path, and choose the path with the highest value. There are eight possible paths:

Path	Total Head Loss
A→H→B→F→E→M→N	1.0 + 1.0 + 2.0 + 1.0 + 3.0 + 1.0 = 9.0
A→H→I→J→G→M→N	1.0 + 0.2 + 0.1 + 0.1 + 0.4 + 1.0 = 2.8
A→H→K→I→J→G→M→N	1.0 + 0.3 + 0.2 + 0.1 + 0.1 + 0.4 + 1.0 = 3.1
A→H→K→L→J→G→M→N	1.0 + 0.3 + 0.3 + 0.2 + 0.1 + 0.4 + 1.0 = 3.3
A→H→K→L→M→N	1.0 + 0.3 + 0.3 + 0.3 + 1.0 = 2.9
A→H→B→D→C→E→M→N	1.0 + 1.0 + 0.2 + 0.1 + 0.2 + 3.0 + 1.0 = 6.5
A→H→B→C→E→M→N	1.0 + 1.0 + 0.2 + 0.2 + 3.0 + 1.0 = 6.4
A→H→B→D→E→M→N	1.0 + 1.0 + 0.2 + 0.3 + 3.0 + 1.0 = 6.5

Therefore, the "free split" is A→H→B→F→E→M→N at a total head loss of 9.0 inches water gage.

Solution 3.4

By following the arrows from the intake to the upper fan, it is obvious that there are four distinctly separate paths that the air can travel. Likewise, there are seven distinctly separate paths that the air can travel between the intake and the lower fan. By adding the pressure drops of each path, and recalling that the sum of the pressure drops around any loop must equal zero, the two free splits can be determined. Thus, the static head for each fan and the regulator placements are shown in the following diagram.

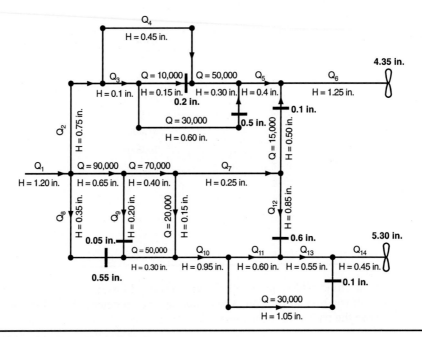

Illustrates Solution to Problem 3.4

Solution 3.5

Using Eq. 3.6,

$$A = 40\left(\frac{Q}{H^{0.5}}\right)$$

$$= 40\left(\frac{0.35}{0.6^{0.5}}\right)$$

$$= 18 \text{ sq ft}$$

Thus, the regulator would have to be constructed so that the opening would be only 18 sq ft.

Solution 3.6

$$R_3 = \frac{KLO}{5.2A^3}$$

for three entries. To determine R_5, the resistance factor for five entries, it is easy to see that (1) K does not change; (2) L does not change; (3) 5.2 is a constant; (4) O, the perimeter, is increased by 5/3; and (5) A, the cross-sectional area of the entries, is increased by 5/3. Therefore,

$$R_5 = \left[\left(\frac{KL}{5.2}\right)\frac{5(O)}{3}\right] / \left[\left(\frac{5}{3}\right)^3 A^3\right]$$

$$R_5 = \frac{1}{(25/9)}\left(\frac{KLO}{5.2A^3}\right) = \left(\frac{9}{25}\right)R_3$$

Thus, the overall resistance is 36% of the three-entry system. This answer can also be verified by Eq. 3.4:

$$\left(\frac{1}{n_{new}/n_{original}}\right)^2 = \left(\frac{1}{5/3}\right)^2 = 0.36$$

Solution 3.7

$$q \propto A\left(\frac{A}{KLO}\right)^{0.5}$$

$$q_a \propto 100\left[\frac{100}{(100 \times 10^{-10})(2{,}000)(50)}\right]^{0.5} = 31{,}623$$

$$q_b \propto 105\left[\frac{105}{(100 \times 10^{-10})(1{,}500)(44)}\right]^{0.5} = 41{,}881$$

$$q_c \propto 72\left[\frac{72}{(100 \times 10^{-10})(1{,}250)(36)}\right]^{0.5} = 28{,}800$$

$$q_d \propto 126\left[\frac{126}{(100 \times 10^{-10})(1{,}900)(50)}\right]^{0.5} = 45{,}887$$

TOTAL: 148,191

$$\frac{31{,}623}{148{,}191} \times 180{,}000 = 38{,}411 \text{ cfm}$$

$$\frac{41{,}881}{148{,}191} \times 180{,}000 = 50{,}871 \text{ cfm}$$

$$\frac{28{,}800}{148{,}191} \times 180{,}000 = 34{,}982 \text{ cfm}$$

$$\frac{45{,}887}{148{,}191} \times 180{,}000 = 55{,}737 \text{ cfm}$$

Solution 3.8

The air traveling along the belt entry cannot be used to ventilate the faces due to federal law. Therefore, only one intake and one return are available.

$$H = RQ^2 = \frac{KSQ^2}{5.2A^3} = \frac{KLOQ^2}{5.2A^3}$$

where:

$A = 5 \times 18 = 90$ sq ft

$O = 18 + 18 + 5 + 5 = 46$ ft

$L = 2{,}500$ ft

intake: $\dfrac{(60 \times 10^{-10})(46)(2{,}500)(40{,}000)^2}{5.2(90)^3} = 0.291$ in.

return: $\dfrac{(80 \times 10^{-10})(46)(2{,}500)(40{,}000)^2}{5.2(90)^3} = 0.388$ in.

$$H_T = 0.679 \text{ in.}$$

$$\text{hp} = \frac{5.2HQ}{33{,}000} = \frac{(5.2)(0.679)(40{,}000)}{33{,}000} = 4.28 \text{ hp}$$

Solution 3.9

To avoid dust suspension, the air velocity in the intake will be limited to 800 fpm. Therefore, the maximum quantity that can be delivered is:

$$Q_T = V_T A_T = (800 \text{ fpm})(4 \text{ ft} \times 15 \text{ ft})$$

$$= 48{,}000 \text{ cfm}$$

The pressure drop along the intake is:

$$H = \frac{KLOQ^2}{5.2A^3} = \frac{(60 \times 10^{-10})(5{,}000)(4 + 4 + 15 + 15)(48{,}000^2)}{5.2(4 \times 15)^3} = 2.34 \text{ in. of water gage}$$

Next, determine how the air will naturally apportion itself between the two returns:

$$q \propto A\left(\frac{A}{KLO}\right)^{0.5}$$

$$q_1 \propto A_1\left(\frac{A_1}{K_1L_1O_1}\right)^{0.5} = 60\left[\frac{60}{(60 \times 10^{-10})(5{,}000)(38)}\right]^{0.5} = 13{,}765$$

$$q_3 \propto A_3\left(\frac{A_3}{K_3L_3O_3}\right)^{0.5} = 48\left[\frac{48}{(100 \times 10^{-10})(5{,}000)(32)}\right]^{0.5} = 8{,}314$$

Thus,

$$Q_1 = \left(\frac{13{,}765}{13{,}765 + 8{,}314}\right)(48{,}000) = 29{,}925 \text{ cfm}$$

$$Q_3 = \left(\frac{8{,}314}{13{,}765 + 8{,}314}\right)(48{,}000) = 18{,}075 \text{ cfm}$$

Obviously, entry 3 has a greater resistance to the flow of air. If a flow of 24,000 cfm is desired for this entry, the pressure drop will be:

$$H = \frac{KLOQ^2}{5.2A^3} = \frac{(100 \times 10^{-10})(5{,}000)(32)(24{,}000^2)}{5.2(48^3)} = 1.60 \text{ in.}$$

The panel air pressure requirement will be 3.94 in. of water gage (2.34 + 1.60 = 3.94), and entry 1 will have to be regulated accordingly.

Solution 3.10

The perimeter (O) for each entry is 44 ft, the cross-sectional area (A) for each entry is 96 sq ft, and the length (L) of each entry is 2,000 ft. The two intakes are interconnected and handle the entire quantity:

$$H = \frac{KLOQ^2}{5.2A^3} = \frac{(55 \times 10^{-10})(44)(2{,}000)(2)(40{,}000)^2}{5.2(96)^3(2)^3} = 0.0421 \text{ in.}$$

The two returns are separated by the intakes and the belt entry. However, because they are in parallel and it is assumed that half the quantity is handled by each, the head loss in one return is equal to the head loss in the other return.

$$H = \frac{KLOQ^2}{5.2A^3} = \frac{(75 \times 10^{-10})(44)(2{,}000)(20{,}000)^2}{5.2(96)^3} = 0.0574 \text{ in.}$$

The total head loss for the panel is merely the summation of the values calculated above.

$$H_T = 0.0421 \text{ in.} + 0.0574 \text{ in.} = 0.0995 \text{ in.}$$

Solution 3.11

The schematic is basically a line diagram of a multiple-entry, unidirectional system of mine ventilation. All of the intakes are in parallel going into any one section, so the head loss in one entry is the same as the head loss in all of the entries; the same can be said for the returns. Also notice that the quantity of air delivered to this end of the mine splits at point A and comes together again at point B. Thus, this layout is identical to the network shown in Figure 3.2. The problem becomes one of unregulated and controlled splitting.

$$q \propto A\left(\frac{A}{KLO}\right)^{0.5}$$

$$q_A = A_A\left(\frac{A_A}{K_A L_A O_A}\right)^{0.5} = 120\left[\frac{120}{((75+95)/2)(7{,}000)(52)}\right]^{0.5} = 0.236$$

$$q_B = 108[108/((2{,}500)(85) + 3{,}500(100))/((6{,}000)(48))]^{0.5} = 0.216$$

Note: The K values have to be altered for the panel because of the difference in lengths of the intakes and returns.

$$q_C = 100[(100)/[(2{,}000(85) + 3{,}000(110))/(5{,}000)](5{,}000)(50)]^{0.5} = 0.200$$

$$0.236 + 0.216 + 0.200 = 0.652$$

$$Q_A = [(0.236)/(0.652)](75{,}000) = 27{,}147 \text{ cfm}$$

$$Q_B = [(0.216)/(0.652)](75{,}000) = 24{,}847 \text{ cfm}$$

$$Q_C = [(0.200)/(0.652)](75{,}000) = 23{,}006 \text{ cfm}$$

Based on the preceding calculations, panel C is the free split. If 25,000 cfm is to be delivered to each panel, then the head loss per panel can be calculated as follows:

$$H = [KLOQ^2]/[5.2A^3]$$

$$H_A = [(85 \times 10^{-10})(7{,}000)(52)(25{,}000^2)]/[5.2(120^3)] = 0.215 \text{ in.}$$

$$H_B = [(93.75 \times 10^{-10})(6{,}000)(48)(25{,}000^2)]/[5.2(108^3)] = 0.258 \text{ in.}$$

$$H_C = [(100 \times 10^{-10})(5{,}000)(50)(25{,}000^2)]/[5.2(100^3)] = 0.300 \text{ in.}$$

Panel A has to be regulated 0.085 in. (0.300 − 0.215), and panel B has to be regulated 0.042 in.

The regulator for panel A should be placed at the end of the return entry just inby point B. The regulator for panel B should be placed at the end of the return entry just inby the overcasts directing the intake to panel A.

Solution 3.12

Plot the mine characteristic curve.

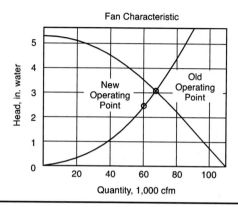

Illustrates Solution to Problem 3.12

From $H = RQ^2$,

$$R = H/Q^2$$

$$R = 2.5/(60,000)^2 = 6.94 \times 10^{-10}$$

Select points to plot the mine characteristic.

$$H = R \times Q^2 = \text{head (in. water gage)}$$

$$H_{(20,000)} = (6.94 \times 10^{-10}) \times (20,000)^2 = 0.28$$

$$H_{(40,000)} = (6.94 \times 10^{-10}) \times (40,000)^2 = 1.11$$

$$H_{(60,000)} = (6.94 \times 10^{-10}) \times (60,000)^2 = 2.50$$

$$H_{(80,000)} = (6.94 \times 10^{-10}) \times (80,000)^2 = 4.44$$

From the characteristic curve, $Q_1 = 67,000$ cfm

From the first fan law,

$$n_2 / n_1 = Q_2/Q_1$$

$$n_2 = (Q_2/Q_1) \, n_1$$

$$n_2 = (60,000/67,000)(800) = 716 \text{ rpm}$$

Solution 3.13

In this problem, the fan is undergoing both a speed and a density change. Thus, the new quantity and head are determined by multiplying the original values by the factors relating to speed and density changes.

$$Q_2 = (n_2/n_1)Q_1$$

$$Q_2 = (716/800)Q_1 = 0.895 Q_1$$

$$H_2 = H_1[(n_2)/n_1]^2[(w_2/w_1)] = H_1(716/800)^2(0.065/0.075) = 0.694 H_1$$

Select points for the new fan characteristic and plot the new fan curve.

Q_1 (cfm)	H_1 (in.)	Q_2 (cfm)	H_2 (in.)
20,000	5.10	17,900	3.54
40,000	4.55	35,800	3.16
60,000	3.60	53,700	2.50
80,000	2.20	71,600	1.53

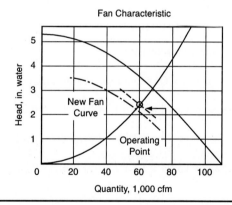

Illustrates Solution to Problem 3.13

With this new fan curve, the fan will produce 58,000 cfm at 2.3 in. of water gage. Therefore, to operate at 60,000 cfm, the required speed change will be:

$$n_2 = n_1 \frac{Q_2}{Q_1} = \frac{716}{(60,000/58,000)} = 741 \text{ rpm}$$

This final speed change will shift the new fan curve up to the 60,000 cfm operating point.

Solution 3.14

Plot the second fan's (fan B) characteristic curve, based on the test results, as indicated on the following page:

Obviously, the new fan will not be able to supply 250,000 cfm at 2.30 in.; however, the two fans may be placed in parallel to achieve the mine's requirements. When fans are placed in parallel, the quantities may be added while keeping the heads constant. The mine characteristic and combined fan curves are shown; the mine characteristic is plotted in the same manner described in problem 3.12, with

$$R = \frac{H}{Q^2} = \frac{2.5}{(250,000)^2} = 0.4 \times 10^{-10}$$

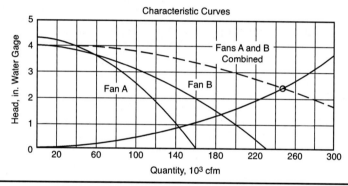

Illustrates Solution to Problem 3.14

From the plot, it can be seen that the fans in parallel will supply 246,000 cfm at 2.40 in. water gage. Therefore, the mine can use the second fan and connect it in parallel with the existing fan. A slight speed change will achieve the exact quantity and head.

Solution 3.15

At the outset, the friction factor for the entries is 49×10^{-10}.

The next step is to divide the mine into airway segments from the collar of the intake air shaft to the face area of 15N and back to the collar of the return air shaft. Linear distances should be noted as well as the number of parallel airways in the segment. In this case, there are 10 segments:

Segments	Length (ft)	Airways
Intake:		
1. Air shaft	1,000	1
2. Air shaft to the base of the bleeders	400	5
3. The base of the bleeders to 14N	6,600	5
4. 14N to 15N	500	5
5. 15N intake	2,000	2
Return:		
1. 15N return	2,000	2
2. 15N to 14N	500	4
3. 14N to the base of the bleeders	6,600	4
4. The base of the bleeders to the air shaft	600	4
5. Air shaft	1,000	1

To determine the total air quantity requirement for the mine, calculate the quantity needed in the north end of the mine and double it to account for the quantity needed in the south end:

20,000	Air quantity to each last open crosscut
× 2	To account for two splits per section
40,000	
+ 20,000	To account for leakage per section
60,000	Air quantity to each section
× 3	To account for the three sections in the north
180,000	
+ 90,000	To account for leakage along Main North
270,000	Total for the northern end of the mine
× 2	To account for the southern end of the mine
540,000	Total for the entire mine

Next, note the air quantity at the beginning, end, and midpoint of the 10 segments determined earlier (assuming air loss is linear):

Segments	Beginning	End	Midpoint
Intake:			
1	540,000	540,000	540,000
2	270,000	235,000	252,500
3	235,000	195,000	215,000
4	135,000	120,000	127,500
5	60,000	40,000	50,000
Return:			
1	40,000	60,000	50,000
2	120,000	135,000	127,500
3	175,000	215,000	195,000
4	235,000	270,000	252,500
5	540,000	540,000	540,000

The northern end of the mine can now be represented by the following line diagram:

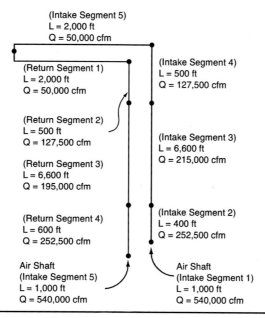

Illustrates Solution to Problem 3.15

Using the equation, $H = KLOQ^2/5.2A^3$, the head loss calculations can be simplified by using the following tabular format. (Note that the values for perimeter, O, and area, A, are based on 6-ft high entries as well as the number of entries in each segment.)

Thus, the mine fan handles 540,000 cfm at a head of 2.25 in. of water gage. Notice that this problem did not take into account the shock losses as they affect the total head of the system. In actual practice, the head may be increased approximately one-third to account for these losses.

Segments	K	L	O	Q	A	H
Intake:						
1	15×10^{-10}	1,000	55	540,000	240	0.335
2	49×10^{-10}	400	220	252,500	480	0.048
3	49×10^{-10}	6,600	220	215,000	480	0.572
4	49×10^{-10}	500	220	127,500	480	0.015
5	49×10^{-10}	2,000	88	50,000	192	0.059
Return:						
1	49×10^{-10}	2,000	88	50,000	192	0.059
2	49×10^{-10}	500	176	127,500	384	0.024
3	49×10^{-10}	6,600	176	195,000	384	0.735
4	49×10^{-10}	600	176	252,500	384	0.112
5	15×10^{-10}	1,000	57	540,000	255	0.289
					Total:	2.25

CHAPTER 4

Strata Control

Perhaps no other area reflects the art as well as the science of mining-engineering design quite as obviously as strata control. This status is due to the fact that it is difficult to design mining structures because rock rarely behaves as a homogeneous medium. In addition, the original state of stress and the mechanical properties of the in situ rock are somewhat difficult to determine. However, the aspects that are hard to treat mathematically can and have been successfully approached using model studies and in situ measurements. This chapter reviews the fundamentals of strata-control design and provides examples of applications to typical problems.

STRESSES AROUND MINE OPENINGS

A close examination of Figure 4.1 reveals that, prior to mining, stresses exerted in the overburden can be in both the horizontal and vertical directions. The example problems in chapter 1 illustrated that the maximum compressive stress (σ_v) at any depth (D) below the surface can be approximated as follows:

$$\sigma_v \text{ (in psi)} = \frac{(62.4)(\text{specific gravity of the overburden})(D, \text{ in ft})}{(144 \text{ sq in. per sq ft})} \quad \text{(EQ 4.1)}$$

Further, the lateral stress (σ_h) is related to the maximum compressive stress by using Poisson's ratio (μ):

$$\sigma_h \text{ (in psi)} = \frac{\mu \sigma_v}{1 - \mu} \quad \text{(EQ 4.2)}$$

As soon as a heading is driven, such as the one in the coal seam shown in Figure 4.2, the stresses shift to the sides, or abutments, of the opening. Clearly, these compressive stresses around the opening exceed the premining conditions. The problem of designing an underground opening, therefore, is to determine the maximum stresses and to note if they exceed the ultimate strength of the rock. Table 4.1 lists several rocks with their corresponding strengths.

Figures 4.3 through 4.5 depict the boundary stress concentrations around circular, ovaloidal, and rectangular openings. M is the ratio of the horizontal stress to the vertical stress or, as defined above, $\mu/(1 - \mu)$. This book concentrates on the state of stress represented by M = 0.33, which corresponds to the condition of no lateral deformation in a

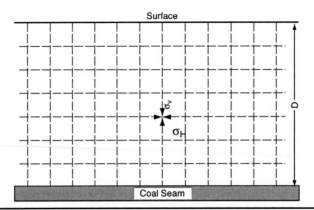

FIGURE 4.1 Original stress distribution on a virgin coal seam

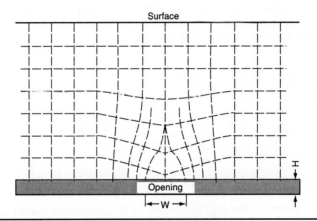

FIGURE 4.2 Stress trajectories around a single opening

rock having a Poisson's ratio of 0.25. For an ovaloid with an opening width-to-height ratio of 2.0 (Figure 4.4), the boundary stress concentrations are at their maximum values at the corners of the opening, with values in excess of 3.0. If the depth of overburden is 500 ft, the maximum boundary tangential stress (σ_t) of the ovaloid is:

$$\sigma_t/\sigma_v \approx 3.0$$

$$\sigma_t \approx 500\ (1.1)(3.0) \approx 1{,}650 \text{ psi}$$

Figure 4.6 relates the maximum tangential stress of the periphery of the opening to the width-to-height ratios of various openings. For the purposes of reducing the maximum tangential stress, it is apparent that an ovaloid is a better choice than an ellipse or a rectangle when the width-to-height ratio exceeds 1.0.

TABLE 4.1 Some properties of rock and coal materials

Rock or Coal Type*	Uniaxial Compressive Strength, psi			Uniaxial Tensile Strength, psi			Modulus of Elasticity, ksi†			Poisson's Ratio		
	From	To	Mean	From	To	Mean	From	To	Mean	From	To	Mean
Basalt	6,090	51,475	21,750	290	4,060	1,885	2,320	14,645	7,685	0.13	0.38	0.22
Dolerite	32,915	46,255	40,600	1,740	3,770	2,900	8,700	13,050	10,150	0.15	0.29	0.20
Gneiss	10,585	49,300	23,055	435	3,045	2,030	2,320	14,935	8,410	0.10	0.40	0.22
Granite	4,350	46,980	24,070	435	5,655	1,740	1,450	10,730	6,525	0.10	0.39	0.23
Limestone	6,960	30,450	14,790	290	5,800	1,740	145	13,340	6,960	0.08	0.39	0.25
Norite	42,050	47,270	43,210	2,175	3,625	2,900	13,050	15,950	14,500	0.21	0.26	0.24
Quartzite	29,000	44,080	36,540	2,465	4,060	3,625	10,150	15,225	13,050	0.11	0.25	0.16
Sandstone	5,800	25,955	13,920	435	1,015	725	1,450	6,670	3,190	0.10	0.40	0.24
Shale	5,220	24,940	13,775	290	725	435	1,450	6,380	4,060	0.10	0.19	0.14
Pittsburgh Coal	2,088	4,307	3,219	276	464	363	218	537	464	—	—	0.37
Pocahontas No. 3 Coal	2,639	2,828	2,741	—	—	—	348	392	377	—	—	—
Herrin No. 6 Coal	1,450	2,045	1,653	—	—	—	450	551	508	—	—	0.42

Source: Bieniawski, 1984, *Rock Mechanics Design in Mining and Tunneling*, A.A. Balkema, Rotterdam.
*Rock specimens were 4.25 in. high and 2.125 in. in diam; coal specimens were 34-in. cubes.
†One ksi equals 1,000 psi.

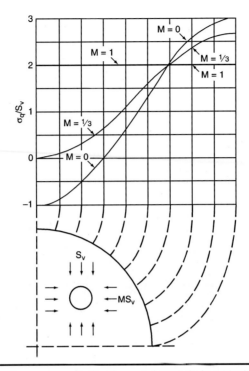

FIGURE 4.3 Boundary stress concentration for a circular hole in a biaxial stress field
Source: Obert et al. 1960.

SAFETY FACTOR

As stated earlier, the state of stress and the mechanical properties of the in situ rock are difficult to determine. To overcome these problems, safety factors are incorporated into the solutions to strata-control problems. The safety factor is defined as the ratio of the breaking stress to the working stress, with recommended values of two to four for pillars and ribs in compression and four to eight for bedded roof in tension.

PILLAR DESIGN

A pillar is that portion of the coal or ore that is left unmined in order to support the overburden. Current design approaches attempt to limit the progressive failure of pillars. To accomplish this, the following information is usually required: (1) premining pillar loading, (2) mining-induced pressures, and (3) pillar strength.

After mining has taken place, each pillar must support the rock column directly above it, as well as that portion of the rock column halfway between it and the neighboring pillars. This is known as the tributary area (see problem 1.13). Obviously, as recovery increases, a greater size of rock column must be supported by each remaining pillar. Although premining pillar loading and mining-induced stress can be estimated fairly readily, the problem in designing pillars revolves around the difficulty of predicting pillar strength.

STRATA CONTROL | 83

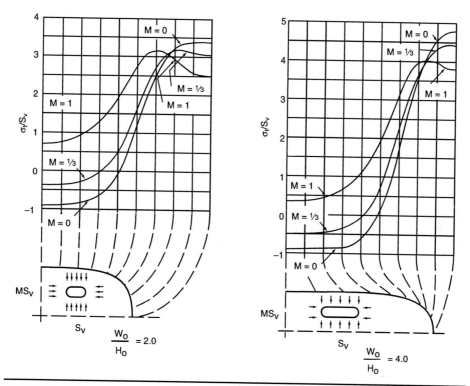

FIGURE 4.4 Boundary stress concentration for ovaloidal holes in a biaxial stress field
Source: Obert et al. 1960.

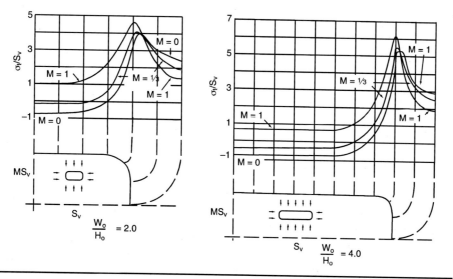

FIGURE 4.5 Boundary stress concentration for rectangular holes with rounded corners
Source: Obert et al. 1960.

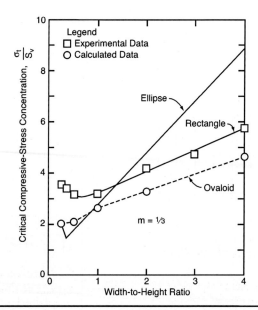

FIGURE 4.6 Critical compressive stress concentrations around openings with various shapes
Source: Obert et al. 1960.

Numerous techniques have been proposed for obtaining adequate pillar strength information at a reasonable cost with relative ease. Coal and noncoal mines have different requirements. The recommended techniques also vary depending on whether the pillars are to be designed solely as support pillars or are to be over-designed for in order to conduct retreat mining. In this text, coal and noncoal pillars are handled separately, but in both instances, attention is directed solely toward the design of support pillars.

Experimental results from tests on rock and coal test specimens indicate that a strength-reduction effect exists with increasing specimen size and that a critical size, defined as that specimen size at which a continued increase in specimen width causes no significant decrease in strength, also exists. For example, it has been reported that 3-ft cubical specimens of coal constitute a critical-size value (Bieniawski, 1992). This concept is significant because the strength values at the critical size are directly applicable to full-sized pillars and the size effect characterizes the difference in strength between small-sized laboratory specimens (σ_1) and the large-sized pillars existing in mines.

Research has shown that the scaling of coal properties measured in the lab to those in the field where cubical pillars have a height >36 in. is as follows:

$$\sigma_1 = \frac{K}{(36)^{0.5}} \qquad \text{(EQ 4.3)}$$

The following should be used where cubical pillars have a height <36 in.:

$$\sigma_1 = \frac{K}{(h)^{0.5}} \qquad \text{(EQ 4.4)}$$

In the above equations, the constant K is a coefficient depending on the particular coal. K is determined from the compression testing of cubical coal specimens and is related in the following manner (Gaddy, 1956):

$$K = S_c(D)^{0.5} \qquad \text{(EQ 4.5)}$$

where S_c is the average ultimate compressive strength of coal cubes tested in pounds per square inch, and D is the edge dimension of the cube in inches. Although there is a difference in laboratory results, depending on whether cylindrical or cubical specimens are used, this difference is, for all practical purposes, not significant when D ranges from 2 to 4 in. (Bieniawski, 1992).

Although numerous pillar-strength formulas have been proposed, a limited number have found wide acceptance for noncoal and coal mines.

In noncoal situations, the Obert and Duvall (1967) equation has been proposed:

$$\sigma_p = C_1\left[0.778 + 0.222\left(\frac{W}{H}\right)\right] \qquad \text{(EQ 4.6)}$$

where σ_p is the pillar strength in pounds per square inch, C_1 is the uniaxial compressive strength of a cubical specimen (w = h) in pounds per square inch, W is the pillar width in inches, and H is the pillar height in inches. In this case, the recommended safety factors are between 2 and 4, with the higher value being associated with long-term requirements.

Pillars in hard-rock mines may also be evaluated using the extraction ratio:

$$R = 1 - \frac{F(S_v)}{C_p} \qquad \text{(EQ 4.7)}$$

where R is the extraction ratio, F is the safety factor, S_v is the average applied stress in pounds per square inch, and C_p is the compressive strength of the pillar in pounds per square inch.

In the United States, research into the design of coal-mine pillars has been conducted for several decades. It has included findings from observations in actual operations, empirical design equations, and numerical design. No matter how precise the numerical design methods may be from a mathematical standpoint, practical coal-mine pillar design presently follows the empirical approach due to its ease of application.

One of the first coal-pillar design equations to meet with acceptance in the United States is the Holland–Gaddy equation. It was developed for coal pillars with width-to-height ratios that vary from 1.0 to 10.0. The equation is as follows:

$$\sigma_p = \frac{K(W)^{0.5}}{H} \qquad \text{(EQ 4.8)}$$

where σ_p is the average ultimate strength of the coal pillar in pounds per square inch, W is the least width of the pillar in inches, and H is the height of the pillar in inches. The recommended safety factor for use with the Holland–Gaddy equation is between 1.8 and 2.2. This equation often results in a conservative estimate of pillar strength.

Bieniawski (1992) proposed the following formula, with a safety factor of 2.0, based on large-scale in situ tests on coal pillars:

$$\sigma_p = \sigma_1\left[0.64 + 0.36\left(\frac{W}{h}\right)\right] \qquad \text{(EQ 4.9)}$$

where σ_p is pillar strength, w is pillar width, h is pillar height, and σ_1 is the strength of a cubical specimen of critical size (≥3 ft). The following modification of Eq. 4.9 was recently proposed for rectangular pillars (Mark, 1999):

$$\sigma_p = \sigma_1[0.64 + 0.54(w/h) - 0.18(w^2/Lh)] \quad \text{(EQ 4.10)}$$

where: L = pillar length.

ROOF-SPAN DESIGN

The design of opening widths is complex and a major concern because most mine fatalities result from falls of the immediate roof. Many different methods have been proposed for the design of opening widths, and all are constrained by federal and state safety regulations. As a result, the design approach may incorporate a combination of concepts.

Where opening widths are critical, such as those in coal mines, beam theories appear to be applicable because of the stratigraphic characteristics of the immediately overlying roof. A simple beam reflects the conditions found with shallow overburden, while a fixed-ends beam is applied to conditions with deep overburden. The opening width for shallow overburden is calculated as follows:

$$L = \left[\frac{4t^2(\sigma_e)}{3W}\right]^{0.5} \quad \text{(EQ 4.11)}$$

where L is the opening width in feet, t is the beam thickness in feet, σ_e is the allowable working stress (modulus of rupture divided by the safety factor) in pounds per square foot, and W is the uniform load per unit length per unit width of the beam. For deep mines, the equation is modified as follows:

$$L = \left[\frac{2t^2(\sigma_e)}{W}\right]^{0.5} \quad \text{(EQ 4.12)}$$

The following general equation has been developed for both of the preceding conditions:

$$L = \frac{4t(\tau_e)}{3W} \quad \text{(EQ 4.13)}$$

where τ_e is the allowable working shear stress (shear strength divided by the safety factor). For the given overburden conditions, if the latter equation yields a lower value of L than the corresponding equation developed by means of the allowable working stress, its value is selected as the maximum allowable roof span without any reinforcement.

Since many operations reinforce the immediate roof of openings with roof bolts, the effects of this activity should be noted. The design of roof-bolting systems incorporates the selection of the roof bolts by diameter and length, spacing, capacity, and tension. Roof bolting aids the stability of the immediate roof by either bonding weak strata into one strong beam or suspending weak strata from a stronger overlying stratum. This, in essence, can alter the span equations that were stated earlier.

When the suspension principle is used, the load per bolt can be determined as follows:

$$P = \frac{wtBL}{(n_1 + 1)(n_2 + 1)} \quad \text{(EQ 4.14)}$$

where P is the load per bolt, w is the unit weight of the roof, t is the beam thickness, B is the span width, L is the depth of the cut, n_1 is the number of rows of bolts, and n_2 is the number of bolts per row.

A roof bolt's yield capacity (C) is normally described in terms of its diameter (D) and the grade of steel (G) (Mark, 2000):

$$C = \left(\frac{B}{4}\right)G(D^2) \qquad \text{(EQ 4.15)}$$

Rebar diameter is usually described by a number that represents the number of eighths of an inch; e.g., #5 rebar is 5/8 in. in diameter and #6 rebar is 6/8 in. (or 3/4 in.) in diameter. The grade of the steel in the bolt is given in thousands of pounds per square inch; e.g., a grade-60 steel is 60,000 psi. A bolt's ultimate capacity is normally greater than the yield. For example, a #6 rebar grade-60 bolt may have a minimum yield of 26,400 lb and a minimum ultimate tensile strength of 39,600 lb (Mark, 2000).

For coal mining applications, 30 CFR 75.300 provides several, important regulations for roof spans. It states that openings should neither exceed 20 ft in width where roof bolting is the sole means of support, nor should they exceed 30 ft when roof bolts and other support, such as timber posts, are used.

It further stipulates that roof bolts should never be less than 30 in. in length and should be anchored at least 1 ft in the stronger strata to suspend the immediate roof. Further, the bolt spacing and the distance between the bolt and the rib or the face should never exceed 5 ft, mechanical bolts should be tensioned to 50% of the yield point of the bolt or the anchorage capacity, and miners may not work under an unsupported roof. Ordinarily, the spacing of roof bolts should satisfy the criterion that, under typical mining conditions, the ratio of bolt length to bolt spacing should be about 1.5, and 2.0 minimum in fractured rock (Bieniawski, 1992).

Even though all of the complexities of roof bolt and ground interactions make it difficult to develop guidelines for roof bolt selection, Mark (2000) recently proposed the following:

1. Evaluate the geology either through underground observation or examination of exploration drill cores. The coal mine roof rating (CMRR) can then be determined. It has been shown that when the CMRR ranges from 65 to 100, roofs are considered strong; when the CMRR ranges from 45 to 65, roofs are considered moderate; and when the CMRR is less than 45, roofs are considered weak. In fact, when the rating is below 37, collapse occurs before a cut can be completed.
2. Evaluate the stress level.
3. Evaluate mining-induced stress.
4. Evaluate the intersection span. The appropriate diagonal intersection span (I_s), in feet, is 31 + (0.66 × CMRR). If the CMRR exceeds 65, then 65 should be used in the equation.
5. Determine the bolt length. For beam building, use the following:

$$L_B = 0.12(I_s)\log_{10}(H)\frac{(100 - \text{CMRR})}{100} \qquad \text{(EQ 4.16)}$$

where L_B is the length of the bolt in feet, I_s is the diagonal intersection span in feet, and H is the depth of cover in feet.

SELECTION OF LONGWALL SHIELDS

The longwall mining method currently accounts for more than 20% of the US annual coal production and approximately 50% of the nation's underground output (Martin, 2000). Its tremendous growth in production output over the past two decades, when the number of faces decreased from 118 faces in 1985 to 62 in 1999, can be attributed to the success of shield-type roof supports. The use of these roof supports introduces another consideration for the strata-control specialist: What is the proper way to select the units?

Currently, the majority of longwall supports are two-leg shields. The average support capacity at yield is 768 tons, with 12 installations employing shields with capacities in excess of 900 tons. Hydraulic setting pressures have remained constant, in the 4,000- to 4,200-psi range. Setting-to-yield ratios are approximately 0.6 (Barczak, 2000).

When designing a shield for a particular application, the following geologic conditions must be understood:

- Composition of the immediate strata, particularly the shelf thickness, that will be carried by the support
- Seam height
- Seam inclination
- Floor conditions

Further, several features of the shield must be defined for a proper application, particularly:

- Area of roof to be supported in square feet
- Tonnage rating at which the shield will yield (This occurs when increasing load from the overlying strata increases the pressure in the shield's legs. At this point, a yield valve set at a predetermined value spills hydraulic fluid to return, allowing the shield to maintain constant resistance against the roof.)
- Tip load in tons (This is the force applied by the shield at the front tip of the canopy.)
- Breakoff load
- Setting-to-yield ratio

To determine the appropriate shield capacity, the following are used to establish the area, volume, and weight of the strata to be supported: mining height (H), caving height (HD, for the worst-case situation), caving angle, roof overhang behind the canopy, web thickness, and shield centers (Figure 4.7). The shield capacity should provide for the worst-case loading situation without reaching yield load. Although shields on 5-ft centers are the most common in current use, recent purchases of shields have been for units averaging 67 in. in width, with designs available for 80 in. (Martin, 2000).

When calculating support resistance, the length of the supported area is the distance taken from the gob break-off line to the coal face to the width of one web thickness; in Figure 4.7, this is represented by W + LK + LS + LB + one web thickness. The width of the supported area is the shield width. Shield capacities provide densities between 8 and 10 tons per sq ft (Martin, 2000).

SUBSIDENCE

When coal is removed by underground mining, the surrounding rock strata tends to slump into the newly created void. Although narrow openings show little, if any, surface

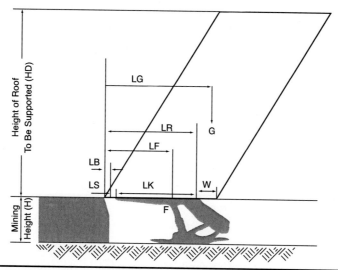

FIGURE 4.7 Dimensions for the determination of shield ratings
Source: Martin 2000.

effect because the remaining coal pillars can support the redistributed pressures, a widening of the opening can cause a weakening of the pillars and subsequent falls of the slumping roof rock. As this widening continues, subsidence goes through a stage of increasing vertical displacement of the surface (subcritical width), to a point where no additional vertical displacement occurs (critical width). Widening the opening beyond this point, usually 1.2 to 1.4 times the depth of overburden, merely increases the maximum subsidence zone in a flattened, trough-like fashion (supercritical width).

To estimate potential damage due to subsidence, a prediction model must be chosen for the evaluation and then strains, due to differential settling of the surface, must be predicted to determine the extent of surface structure damage. Recent work indicates that the hyperbolic tangent function is applicable to the Appalachian bituminous coalfields as a predictor for the subsidence profile:

$$S(x) = (0.5)(S_{max})\left(1 - \tanh\frac{cx}{b}\right) \qquad \text{(EQ 4.17)}$$

where $S(x)$ is the surface subsidence for various distances from the rib of a given opening (for example, above a 1,000-ft longwall face when it is assumed that the point of inflection is directly above the rib of the opening, x equals 0 above the rib line, x equals –500 above the panel center, and x takes on positive values as you move away from the panel); S_{max} is the maximum subsidence observed over a given opening; b is the distance from the inflexion point, or point of maximum slope, to the location where the curve flattens out at maximum subsidence (if, for example, the point of inflexion is directly above the edge of the opening in a 1,000-ft longwall face, b becomes 500 because that is the distance from the rib of the opening to the middle of the face); and c is a constant representing a characteristic of the type of subsidence that occurred, such as subcritical (value of 1.4), critical, or supercritical (values of 1.8 to 2.0) (Peng, 1978).

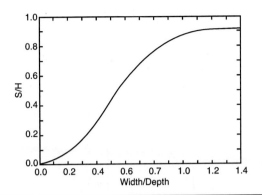

FIGURE 4.8 Subsidence development curve
Source: Peng, *Coal Mine Ground Control*, 1978.

Although subsidence-control research has expanded in the United States during recent years, past references to damage control have been based on the findings of European strata-control specialists. The United Kingdom, in particular, has been prominent in subsidence research. Its National Coal Board (NCB) published the *Subsidence Engineer's Handbook* in 1966 and revised it in 1975 (NCB, 1975). In a study of more than 150 coal mines with depths of overburden ranging from 100 to 2,600 ft, seam thicknesses ranging from 2 to 18 ft, and opening widths ranging from 100 to 1,500 ft, the NCB found that if the ratio of the opening width (W) to the depth of cover (D) was maintained below 0.25, subsidence damage to any type of surface structure was negligible. A graphic portrayal of the NCB findings, with the width-to-depth ratio along the horizontal axis and the vertical surface displacement-to-seam thickness ratio along the vertical axis, is shown in Figure 4.8.

The procedure for subsidence prediction, based on the NCB approach, is as follows:

1. Calculate the width-to-depth ratio and determine the subsidence factor (S/H) using Figure 4.8.
2. Calculate the maximum anticipated subsidence, which is equal to (S/H) (seam height).
3. Determine the transverse subsidence profile using values from Table 4.2.
4. Determine the transverse strain profile, indicated by S_{max}/D.
5. Using Figure 4.9, determine the maximum extension and compression with respect to S_{max}/D.
6. Assess the surface damage by locating the structure with respect to the subsidence profile, determine the maximum strain under the structure using the values from Table 4.3, compare the value with the ranges established by the NCB in Figure 4.10, and determine the subsidence conditions at the building.

Note that the values from the NCB-derived approach assume a 35° angle of draw and that, in the United States, typical values for angles of draw are normally lower.

Control of the damaging effects of subsidence on room-and-pillar mining in the United States has differed from the steps taken for longwall mining. Since room-and-pillar mining limits opening widths, the focus has been on controlling the angle of

STRATA CONTROL | 91

TABLE 4.2   Relationship between W/D and d/D for various points on a subsidence profile

s/S values	0.0	0.10	0.20	0.30	0.40	0.50	0.60	0.70	0.80	0.90	1.00
W/D				Distances from Panel Center in Terms of Depth							
2.60	2.00	1.39	1.29	1.24	1.19	1.16	1.12	1.08	1.03	0.95	0.41
2.40	1.90	1.29	1.19	1.14	1.10	1.06	1.02	0.98	0.93	0.85	0.31
2.20	1.80	1.19	1.09	1.04	1.00	0.96	0.92	0.88	0.83	0.75	0.23
2.00	1.70	1.09	0.99	0.94	0.90	0.86	0.82	0.78	0.73	0.65	0.16
1.80	1.60	1.00	0.90	0.84	0.80	0.76	0.72	0.68	0.63	0.55	0.10
1.60	1.50	0.90	0.80	0.74	0.70	0.66	0.62	0.58	0.63	0.45	0.05
1.40	1.40	0.80	0.70	0.64	0.60	0.56	0.52	0.48	0.43	0.35	0.01
1.30	1.35	0.75	0.65	0.59	0.55	0.51	0.47	0.43	0.38	0.30	0.00
1.20	1.30	0.70	0.60	0.54	0.50	0.46	0.42	0.38	0.33	0.25	0.00
1.10	1.25	0.65	0.55	0.50	0.45	0.42	0.38	0.34	0.29	0.21	0.00
1.00	1.20	0.61	0.51	0.45	0.41	0.37	0.33	0.29	0.24	0.18	0.00
0.90	1.15	0.57	0.46	0.40	0.36	0.32	0.29	0.25	0.20	0.14	0.00
0.80	1.10	0.52	0.42	0.36	0.32	0.28	0.25	0.21	0.17	0.11	0.00
0.70	1.05	0.49	0.39	0.33	0.29	0.25	0.21	0.18	0.14	0.10	0.00
0.60	1.00	0.47	0.36	0.30	0.26	0.22	0.19	0.16	0.13	0.09	0.00
0.50	0.95	0.47	0.34	0.28	0.24	0.21	0.17	0.15	0.12	0.08	0.00
0.40	0.90	0.47	0.34	0.28	0.24	0.21	0.18	0.15	0.12	0.08	0.00
0.30	0.85	0.50	0.38	0.32	0.27	0.23	0.20	0.17	0.13	0.09	0.00
0.20	0.80	0.57	0.48	0.41	0.37	0.32	0.28	0.23	0.19	0.13	0.00

Source: Peng, *Coal Mine Ground Control*, 1978.

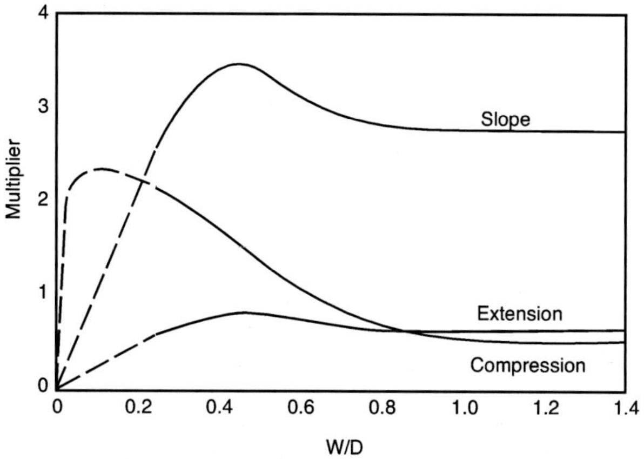

FIGURE 4.9   Graph for predicting maximum slope and strain for various width-to-depth ratios of a panel
Source: from Peng, *Coal Mine Ground Control*, 1978.

TABLE 4.3 Relationship for various strain values in a subsidence profile

	Extension (+E)								Compression (−E)									
	e/E Values:																	
W/D	0.00	0.20	0.40	0.60	0.80	1.00	0.80	0.00	0.20	0.40	0.60	0.80	1.00	0.80	0.60	0.40	0.20	0.00
	Distances from Panel Center in Terms of Depth																	
2.60	2.00	1.58	1.47	1.41	1.36	1.30	1.26	1.16	1.14	1.11	1.08	1.05	0.99	0.90	0.83	0.77	0.70	0.50
2.20	1.80	1.38	1.27	1.21	1.15	1.10	1.06	0.96	0.94	0.91	0.88	0.85	0.79	0.70	0.63	0.57	0.50	0.30
2.00	1.70	1.28	1.17	1.11	1.05	1.00	0.96	0.86	0.84	0.81	0.78	0.75	0.69	0.60	0.53	0.47	0.40	0.20
1.80	1.60	1.17	1.07	1.01	0.95	0.90	0.86	0.76	0.73	0.71	0.68	0.65	0.59	0.50	0.43	0.37	0.30	0.10
1.60	1.50	1.08	0.97	0.91	0.85	0.80	0.76	0.66	0.63	0.61	0.58	0.55	0.49	0.40	0.33	0.27	0.20	0.03
1.40	1.40	0.98	0.87	0.81	0.75	0.70	0.66	0.56	0.53	0.51	0.48	0.45	0.39	0.30	0.23	0.17	0.10	0.00
1.30	1.35	0.93	0.82	0.76	0.70	0.65	0.61	0.51	0.49	0.46	0.43	0.40	0.34	0.25	0.18	0.12	0.05	0.00
1.20	1.30	0.88	0.77	0.71	0.66	0.61	0.56	0.46	0.44	0.41	0.38	0.35	0.29	0.20	0.13	0.07	0.02	0.00
1.10	1.25	0.83	0.72	0.66	0.61	0.56	0.52	0.42	0.39	0.37	0.33	0.31	0.24	0.15	0.09	0.03	0.00	0.00
1.00	1.20	0.79	0.68	0.62	0.57	0.51	0.47	0.37	0.35	0.32	0.29	0.26	0.20	0.10	0.05	0.00	0.00	0.00
0.90	1.15	0.74	0.63	0.57	0.52	0.46	0.42	0.33	0.30	0.28	0.24	0.21	0.15	0.06	0.02	0.00	0.00	0.00
0.80	1.10	0.69	0.58	0.53	0.48	0.42	0.37	0.29	0.26	0.24	0.20	0.17	0.11	0.02	0.00	0.00	0.00	0.00
0.70	1.05	0.65	0.54	0.48	0.44	0.37	0.33	0.25	0.23	0.20	0.17	0.14	0.00	0.00	0.00	0.00	0.00	0.00
0.60	1.00	0.62	0.52	0.45	0.40	0.34	0.29	0.22	0.20	0.18	0.14	0.11	0.00	0.00	0.00	0.00	0.00	0.00
0.50	0.95	0.60	0.51	0.43	0.38	0.32	0.27	0.21	0.18	0.16	0.12	0.08	0.00	0.00	0.00	0.00	0.00	0.00
0.40	0.90	0.61	0.52	0.45	0.40	0.32	0.28	0.21	0.18	0.15	0.11	0.07	0.00	0.00	0.00	0.00	0.00	0.00
0.30	0.85	0.65	0.57	0.51	0.47	0.37	0.32	0.23	0.20	0.16	0.12	0.08	0.00	0.00	0.00	0.00	0.00	0.00
0.20	0.80	0.74	0.69	0.66	0.61	0.49	0.42	0.32	0.27	0.23	0.18	0.12	0.00	0.00	0.00	0.00	0.00	0.00

Source: Peng, *Coal Mine Ground Control*, 1978.

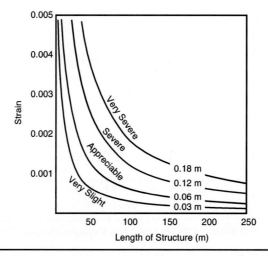

FIGURE 4.10 Relationship of damage to length of structure and horizontal ground strain
Source: Peng, *Coal Mine Ground Control*, 1978.

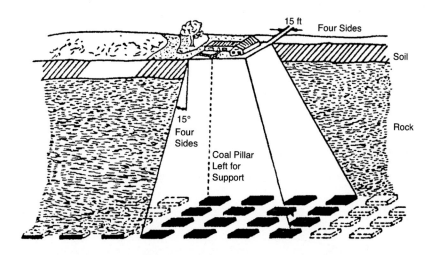

FIGURE 4.11 Protection practice of surface structures in Pennsylvania
Source: Peng, *Coal Mine Ground Control,* 1978.

draw. For example, the Pennsylvania Bituminous Mine Subsidence Act of 1966, the first comprehensive subsidence law passed by any state, provided protection for certain structures that were in place on the effective date of the act. In defining the support area, the angle of draw was assumed to be 15°, thereby affecting the guidelines in the following manner:

> Where a structure is located on terrain that is level or slopes equal to or less than 5%, the lateral distance (LD) of the support area from each side of the structure is equal to the tangent of 15° (0.27), multiplied by the depth of cover (D), plus a safety factor of 15 ft, or:
>
> $$LD = (0.27D) + 15 \qquad \text{(EQ 4.18)}$$
>
> To find the total length or width of the support area, double the result found for LD and add the length or width of the structure. Structures located on terrain that slopes more than 5% on one or more sides require that additional lateral distance (ALD) be added to the downhill side or sides. This additional lateral distance is equal to the percent of slope multiplied by the depth of cover, or:
>
> $$ALD = (\text{percent of slope})(D) \qquad \text{(EQ 4.19)}$$
>
> The above product is added to the downhill side of the support area. (Peng 1978)

The Pennsylvania guidelines dealt solely with defining the support area. Within the area, 50% of the coal had to be left in place in uniformly distributed pillars, which could not be smaller than 20 by 30 ft. Further, no mining could be conducted where the overburden was less than 100 ft under a protected structure, and pillars could not be extracted between two support areas where the distance between them was less than the cover. Figure 4.11 provides a pictorial representation of the guidelines as they have been applied to the protection of a residential home on level terrain. Over the

94 | MINING ENGINEERING ANALYSIS

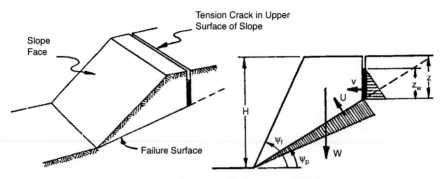

Geometry of Slope with Tension Crack in Upper Slope Surface

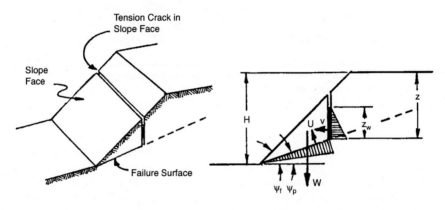

Geometry of Slope with Tension Crack in Slope Face

FIGURE 4.12 Planar failure
Source: Hoek and Bray 1977.

years, Pennsylvania has modified the manner in which the guidelines have been enforced to reflect the changing nature of subsidence control, such as the protection of perennial streams.

SLOPE STABILITY

A significant strata-control problem that an open-pit mine design engineer must eventually confront deals with slope stability. Many factors, such as geology of the area and of the slope material, topography, climate, and hydrology, can influence slope stability. The only parameters that are controllable by the engineer are slope height and angle, standing time, and, to some extent, groundwater level in the pit wall. To determine the stability of a given pit wall, slide mechanics is often used.

To conduct a slope-stability analysis, an attempt is usually made to find an expression for the safety factor (F) of a given failure surface; usually a safety factor of 1.3 is considered adequate for slopes that will remain standing for only a short time, whereas a value of 1.5 is desirable for long-term stability. Although there are four common methods of slope failure (planar, wedge, circular, and toppling), the following discussion will use planar failure as an example because it is the most common form found in coal and hard-rock mining.

Planar failure results when a competent block of rock lies along a planar discontinuity that dips with regard to the slope (Figure 4.12). When this dip angle is greater than the peak friction angle for the discontinuity surface, the block will tend to slide into the pit. Referring once again to Figure 4.12, there are often tension cracks that can fill with water. The presence of water in these cracks can be detrimental because of the buoyant effect that water can have on the potential failure surface. It reduces the effective normal force on the surface and, ultimately, the available friction to oppose sliding along the plane. As such, the safety factor for planar failure is approximately:

$$F = cA + \frac{[W(\cos\psi_p) - U - V(\sin\psi_p)]\tan\phi}{W(\sin\psi_p) + V(\cos\psi_p)} \quad \text{(EQ 4.20)}$$

where (with reference to Figure 4.12) c is the cohesion, A is the length of the slip surface and is equal to $(H - z) \operatorname{cosec}(\psi_p)$, U is the uplift due to water pressure and is equal to 0.5 $(\gamma_w)(z_w)(H - z)(\operatorname{cosec}\psi_p)$, V is the force from the water in the tension crack and is equal to 0.5 $(\gamma_w)(z_w^2)(\sin\psi_p)$, W is the weight of the block, ψ_p is the angle of the discontinuity plane, ϕ is the friction angle, and γ_w is the density of water (62.4 lb per cu ft).

REFERENCES

Adler, L., and M. Sun. 1976. *Ground Control in Bedded Formations*. Research Division Bulletin 28, Blacksburg, VA: Virginia Polytechnic Institute and State University.

Barczak, T. 2000. Examining longwall shield failures from an engineering design and operational perspective. *Proceedings: New Technology for Coal Mine Roof Support*. IC 9453, NIOSH–Pittsburgh Research Laboratory, Pittsburgh, PA, pp. 223–243.

Bieniawski, Z.T. 1984. *Rock Mechanics Design in Mining and Tunneling*. Rotterdam: A.A. Balkema.

———. 1992. Ground Control. In *SME Mining Engineering Handbook*. Littleton, CO: Society for Mining, Metallurgy, and Exploration, Inc.

Gaddy, F.L. 1956. A Study of the Ultimate Strength of Coal as Related to the Absolute Size of the Cubical Specimens Tested. In *Engineering Experiment Station Bulletin*. Blacksburg, VA: Virginia Polytechnic Institute and State University.

Hoek, E., and J. Bray. 1977. *Rock Slope Engineering*. London, England: The Institution of Mining and Metallurgy.

Logie, C.V., and G.M. Matheson. 1982. A Critical Review of the Current State-of-the-Art Design of Mine Pillars. In *First International Congress on Stability in Underground Mining*. Vancouver, British Columbia, Canada, pp. 359–381.

Mark, C. 1999. Empirical Methods for Coal Pillar Design. *Proceedings of the Second International Workshop on Coal Pillar Mechanics and Design*. IC 9448, NIOSH–Pittsburgh Research Laboratory, Pittsburgh, PA, pp. 145–154.

———. 2000. Design of Roof Bolt Systems. *Proceedings: New Technology for Coal Mine Roof Support*. IC 9453, NIOSH–Pittsburgh Research Laboratory, Pittsburgh, PA, pp. 111–131.

Martin, H. 2000. Personal Communication: DBT America Inc.

NCB. 1975. *Subsidence Engineers' Handbook*. London, England: National Coal Board, Mining Department.

Obert, L., and W.I. Duvall. 1967. *Rock Mechanics and the Design of Structures in Rock*. New York: John Wiley and Sons, Inc.

Obert, L., W.I. Duvall, and R.H. Merrill. 1960. *Design of Underground Openings in Competent Rock*. Bulletin 587, Washington, DC: US Bureau of Mines.

Peacock, A. 1981. Design of Shield Supports for the US Coal Mining Industry. In *First Conference on Ground Control in Mining*. Morgantown, WV, pp. 174–185.

Peng, S.S. 1978. *Coal Mine Ground Control*. New York: John Wiley and Sons, Inc.

Singh, M.M. 1992. Mine Subsidence. In *SME Mining Engineering Handbook*. Littleton, CO: Society for Mining, Metallurgy, and Exploration, Inc.

PROBLEMS

Problem 4.1

A 15-ft-wide entry is driven by a full-face boring-type continuous miner in a 5-ft seam at a depth of 500 ft. What is the maximum compressive stress created around the opening?

Problem 4.2

A circular tunnel is to be driven through a competent, homogeneous deposit of sandstone (assume that $\mu = 0$). Assume that the vertical stress gradient in the rock mass is 1.1 psi per ft of cover, the design requires long-term safety factors, and the rock has the following in situ properties:

Compressive strength: $C_o = 24{,}000$ psi

Tensile strength: $T_o = 1{,}500$ psi

What is the maximum depth possible?

Problem 4.3

Assume that a tunnel is to be constructed 600 ft below the surface, as shown. The properties of the rock through which it is to be driven are tabulated as follows:

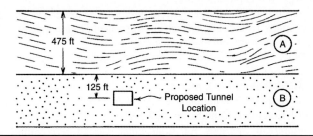

Illustrates Problem 4.3

Strata	Specific Gravity	μ	C_o (psi)	T_o (psi)
A	2.35	0.33	4,800	795
B	2.59	0.25	13,150	1,225

Calculate the stress concentration factors occurring at the center top, corner, and center sidewall for rectangular openings with width-to-height ratios of 2.0 and 4.0. Calculate the vertical and horizontal stresses at the proposed tunnel site. Calculate the critical stresses associated with the various-shaped rectangular openings. Calculate the safety factors for the various critical stresses, and indicate which openings will have long-term stability and which will have short-term stability.

Problem 4.4

A 5-ft-thick coal seam occurs 400 ft below the surface. The rock overlying the coal is nearly horizontal and of such density that it may be assumed that the vertical stress gradient is approximately 1.2 psi per foot of depth. Laboratory tests on the coal indicate that its compressive strength is 3,000 psi. Assume that the laboratory strength data must be derated by a factor of 2 for use in underground pillar design. Assuming a safety factor of 1.0, what is the maximum extraction ratio that can be achieved during development?

Problem 4.5

A coal mine's entry and breakthrough centers are set at 60 ft with 20-ft-wide openings. The 6-ft coal seam lies 1,000 ft below the surface, and the specific gravity of the overburden is 2.4. If the compressive strength of 4-in. cubic coal specimens is 5,000 psi, determine the safety factor for compressive failure of the pillars utilizing the Holland–Gaddy relationship.

Problem 4.6

A lead-zinc mine is to be developed by a room-and-pillar layout. The mine is located at a depth of 500 ft, the overlying rock has an average specific gravity of 2.58, and the pillar rock was tested and found to have a uniaxial compressive strength of 15,000 psi. What safety factor will be necessary to allow an extraction ratio of 75%?

Problem 4.7

A 2-in. cube coal specimen, which was tested in a lab, failed at 8,000 lb. A set of main entries is to be designed through the 5-ft seam from which the specimen was extracted. If there is 500 ft of overburden, and assuming a 50% recovery, what will be the least pillar dimension?

Problem 4.8

A 6-ft coal seam at a depth of 750 ft is to be mined by a room-and-pillar system incorporating 16-ft entries and pillars that are 50 ft square. The K value for the seam is 4,000. Utilizing the Bieniawski formula, determine if these dimensions provide adequate stability.

Problem 4.9

A 6-ft seam is located at a depth of 650 ft. In the preliminary stages of mine design, it is desired that the entries and crosscuts be driven equal to the maximum width allowed by federal law for one means of support. What is the minimum pillar width if uniformly distributed square pillars are to be designed at a desired recovery of 50%? If the average compressive strength of a sufficiently large sampling of 4-in. cube coal specimens taken from the seam was determined to be 2,826 psi, will the proposed pillars be large enough for stability? Use the Holland–Gaddy formula to verify the pillar size and incorporate a safety factor of 2.0.

Problem 4.10

A 12-ft-thick limestone bed is located 600 ft below the surface. Testing of a specimen whose width-to-height ratio is 0.5 indicates that the limestone has a strength of 9,000 psi in compression. The density of the overburden is 160 lb per cu ft, and indications show that 20-ft wide entries may be driven. A pillar safety factor of 2.5 is to be incorporated. Calculate the pillar dimensions and the initial extraction ratio.

Problem 4.11

A coal mine operation is planning to use untensioned resin bolts to support 20-ft-wide entries driven 20 ft per cut in a 6-ft seam. Management would like to use the fewest number of bolts required by federal law. If beam theory leads management to believe that a competent 5-ft-thick beam can be created in the immediate roof, and the maximum allowable load on the bolts is 80% of yield, would a #6 rebar grade-60 bolt (0.75 in. diameter, 60,000 psi yield strength) be sufficient? Assume that sedimentary rock weighs 160 lb per cu ft, and that bolting can get no closer than 2 ft outby the face.

Problem 4.12

The immediate roof in a new coal-mining section under shallow cover has been found to be free of any major defects. Its thickness is 2 ft, its modulus of rupture is 625 psi, and its specific gravity is 2.4. Utilizing a safety factor of 8, what is the maximum safe roof span in feet?

Problem 4.13

Depicted below is a geologic column of the immediate roof for a coal seam located 600 ft below the surface that will be deep mined. Determine an appropriate bolt length if the roof is considered to have a CMRR of 45 and the entries and crosscuts will be driven 18 ft wide.

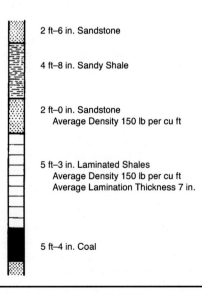

Illustrates Problem 4.13

Problem 4.14

A 4-ft-thick layer of moderately fractured shale lying beneath a massive bed of limestone is to be supported above a 16-ft-wide entry. Each lift will be 20 ft long, the unit weight of the shale is 160 lb per cu ft, and the bolts will be installed on 4-ft centers. Assuming a safety factor of 2.0, determine the bolt load and the diameter of the bolts if grade-40 steel is used.

Problem 4.15

A mine entry that has been exhibiting stability problems is to be monitored for roof sag, bed separation, and floor heave. A set of free-floating roof bolts is to be installed as shown below. After 1 week, the following measurements were made between bolt heads:

Extensometer Readings	T = 0	T = 1 week
$L_{A,B}$	50.3 in.	46.1 in.
$L_{C,D}$	52.3 in.	46.3 in.
$L_{E,F}$	49.1 in.	46.5 in.

Determine the sag of the immediate roof, the heave of the immediate floor, and the bed separation between the immediate roof and the main roof.

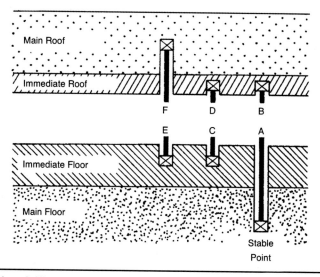

Illustrates Problem 4.15

Problem 4.16

A two-leg shield support (unit width: 5 ft; roof-beam length: 11 ft; length of area to be supported: 14 ft) is designed for a support density of 10 tons per sq ft. If the immediate roof that it must handle weighs 160 lb per cu ft and the induced cleavage planes are oriented in a desirable direction at an angle of 30°, can it handle a shelf thickness (HD) of 10 times its maximum unit height (7.17 ft)?

Problem 4.17

Using the National Coal Board's subsidence development curve, what would be the maximum anticipated subsidence (in feet) of the surface overlying a 6-ft coal seam at a depth of 1,000 ft? The face length is 800 ft and substantial chain pillars are to be left between the longwalls to act as barrier pillars.

Problem 4.18

A pipeline that must be protected lies below the surface and above a 5-ft coal seam located at a depth of 1,200 ft. Using the National Coal Board's subsidence development curve, determine the maximum permissible width of a longwall face if the maximum vertical movement permitted on the surface is 2 ft?

Problem 4.19

A 30- × 100-ft church in southwestern Pennsylvania is located where the surface slopes 3% parallel to the shorter side. Below it, at a depth of 600 ft, lies a 5-ft coal seam that is to be mined using the room-and-pillar technique with 18-ft-wide entries. Determine the dimensions of the support area to limit subsidence-induced damage that has been practiced for many years in Pennsylvania.

Problem 4.20

Suppose that a mine in Pennsylvania owns the mineral rights under a building that has to be protected. The calculated support area is 350- × 450-ft. Mine management has a choice of buying the structure for $30,000, which will enable an extraction ratio of 0.8, or not purchasing the structure and mining only 50% of the 6-ft coal seam. If the company averages a profit of $3.50 per ton of coal and the coal has an average density of 80 lb per cu ft, should the mine buy the building?

Problem 4.21

A 6-ft coal seam is located at a depth of cover of approximately 1,000 ft. If the angle of draw for the overburden is assumed to be 25° and the critical width is estimated to be 2[tan(angle of draw)] (depth of cover), determine the critical width of an opening in the seam with regard to subsidence. How does the calculated value compare to the predicted value developed for the National Coal Board's *Subsidence Engineers' Handbook*?

Problem 4.22

A 1,000-ft-wide, 20,000-ft-long longwall panel, at an average depth of 1,430 ft, is to be driven in a 6-ft coal seam. The surface terrain is fairly flat, but there is considerable concern for subsidence damage to a 50- × 100-ft structure located approximately midpanel directly over the edge of the projected opening. The 50-ft dimension of the structure is parallel to the direction of mining.

Using the NCB approach, determine the maximum subsidence anticipated, the transverse subsidence profile, the transverse strain profile, and the maximum strain developed across the structure. What kind of damage should be expected at the structure?

Problem 4.23

Based on the calculations from problem 4.22, plot the points of the panel's subsidence profile on the diagram shown below. Note that it is arranged to show the right-hand side of the panel, with the units for the horizontal and vertical axes, as shown.

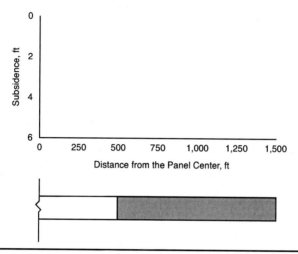

Illustrates Problem 4.23

Problem 4.24

As a basis for comparison, take the information provided in problem 4.22 and recalculate the points along the subsidence curve based on the hyperbolic tangent function. Assume that the maximum subsidence is the same. Plot this profile on the same diagram used for the NCB-derived profile.

Problem 4.25

A quarry face, cut to an angle of 60° with the horizontal, has been excavated to a depth of 100 ft, as shown in the figure on the next page. At this depth, the quarry floor intersects a well-developed joint that dips at an angle of 30° toward the quarry face. This joint is intersected by a vertical tension crack 20 ft back from the quarry face. If the density of the rock is 160 lb per cu ft, the friction angle is 25°, and the cohesion along the joint plane is 1,000 psf, determine the existing factor of safety against sliding of the block. Assume that the situation is not influenced by water pressure.

Problem 4.26

What would be the factor of safety in the previous problem if water fills the vertical tension crack within 20 ft of the top of the quarry bench?

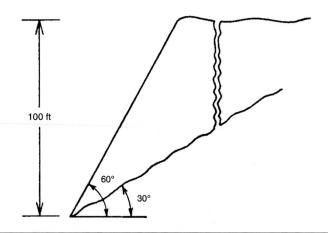

Illustrates Problem 4.25

PROBLEM SOLUTIONS

Solution 4.1

The width-to-height ratio of the opening is 3.0(15/5). Using Figure 4.6 and noting that a boring machine drives an ovaloidal opening, the critical compressive-stress concentration corresponding to a width-to-height ratio of 3.0 is about 4.0. If the vertical stress is assumed to be 1.1 psi per foot of overburden, the maximum compressive stress can be calculated as follows:

$$\sigma_v = (4.0)(500)(1.1) = 2{,}200 \text{ psi}$$

Solution 4.2

$$M = \mu/(1-\mu) = 0/1 = 0$$

Utilizing Figure 4.3, the maximum stress concentration factor at the top of the opening is −1.0 (tension), while the maximum stress concentration factor at the side of the opening is +3.0 (compression). For long-term stability (safety factors: 8 for tension, 4 for compression), the maximum tensile stress is:

$$\sigma_{\tau \, max} = (1{,}500 \text{ psi})/8 = 187.5 \text{ psi}$$

and the maximum compressive stress is:

$$\sigma_{c \, max} = 24{,}000 \text{ psi}/4 = 6{,}000 \text{ psi}$$

Now, due to the stress concentration factor:

$$\text{tensile: } S_v = \text{SCF}(\sigma_{\tau \, max}) = |-1.0\,(187.5)| = 187.5 \text{ psi}$$

compressive: $S_v = \text{SCF}(\sigma_{c\,max}) = |3(6,000)| = 18,000$ psi

Thus, the tensile critical depth is:

$$S_v = 1.1\,h = 187.5$$

$$h = 170 \text{ ft}$$

and the compressive critical depth is:

$$S_v = 1.1\,h = 18,000 \text{ psi}$$

$$h = 16,364 \text{ ft}$$

Therefore, due to tensile considerations, the maximum depth possible is 170 ft.

Solution 4.3

Figure 4.5 can be used for determining the stress concentration factors:

At the tunnel site, $M = \mu/(1 - \mu) = 0.25/(1 - 0.25) = 1/3 = 0.33$.

W_o/H_o	Center (top)	Corner	Center (side)
2	−0.20	4.00	2.00
4	−0.40	5.50	2.60

$$S_v = (\gamma)(h)$$

where γ = (specific gravity)(62.4 lb per cu ft)

Due to strata A: $|\,2.35\,(62.4)(475)/144\,| = 483.7$ psi

Due to strata B: $|\,2.59\,(62.4)(125)/144\,| = 140.3$ psi

$$S_v = 624.0 \text{ psi}$$

$$S_h = \frac{\mu(S_v)}{1-\mu} = \frac{0.25(S_v)}{0.75} = 0.33 S_v = 208 \text{ psi}$$

Critical Stresses

W_o/H_o	$\sigma_{tensile} = |S_{vx}(\text{SCF}_{min})|$	$\sigma_{compressive} = |S_{vx}(\text{SCF}_{max})|$
2.00	$|(0.20)(624)| = 125$ psi	$|(-4.0)(-624)| = 2,496$ psi
4.00	$|(0.40)(624)| = 250$ psi	$|(-5.50)(-624)| = 3,432$ psi

Safety factors to be used:

	Short-term	Long-term
Compressive members	2	4
Tensile members	4	8

W_o/H_o	Compression SF = $\|C_o\|/\|S_c\|$	Stability		Tension SF = $\|T_o\|/\|S_t\|$		
		Short-term	Long-term		Short-term	Long-term
2	(13,150 / 2,496) = 5.27	Yes	Yes	(1,225 / 125) = 9.8	Yes	Yes
4	(13,150 / 3,432) = 3.83	Yes	No	(1,225 / 250) = 4.9	Yes	No

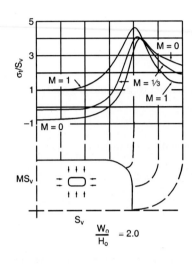
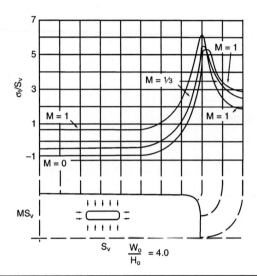

Illustrates Solution to Problem 4.3

Solution 4.4

The vertical stress, σ_v, is:

$$\sigma_v = 1.2D = (1.2 \text{ psi per foot of depth})(400 \text{ ft}) = 480 \text{ psi}$$

Pillar strength, after derating of the cube coal specimen, is:

$$S_{pc} = \frac{Sc}{2} = \frac{3{,}000 \text{ psi}}{2} = 1{,}500 \text{ psi}$$

To achieve stability, the pillar strength must be either greater than or equal to the pillar stress, or:

$$S_{pc} > \sigma_v \frac{1}{(1-R)} = 480 \frac{1}{(1-R)}$$

Equating S_{pc} to $[480/(1-R)]$, and rearranging:

$$R = \frac{1{,}500 - 480}{1{,}500} = 68\%$$

Solution 4.5

Utilizing the Holland–Gaddy relationship, the ultimate compressive strength of the pillar can be determined:

$$K = \sigma_c(D)^{0.5} = 5{,}000(4)^{0.5} = 10{,}000$$

$$\sigma_p = \frac{K(W)^{0.5}}{H} = \frac{(10{,}000)(40)(12)^{0.5}}{(6)(12)} = 3{,}043 \text{ psi}$$

$$F = \frac{3{,}043 \text{ psi}}{\text{average pillar stress}}$$

average pillar stress = (premining stress)(tributary area/pillar area)

$$= (1{,}000 \text{ ft})\left(\frac{(2.4)(62.4)}{144}\right)\left(\frac{3{,}600}{1{,}600}\right) = 2{,}340 \text{ psi}$$

Therefore, F = (3,043 psi/2,340 psi) = 1.3

As the recommended range for the safety factors to be used with the Holland–Gaddy relationship are 1.8 to 2.2, the calculated safety factor is low.

Solution 4.6

Rearranging Eq. 4.7 yields the following relationship in terms of the safety factor:

$$F = (1 - R)(C_p / S_v)$$

Since the vertical compressive stress is 559 psi, [(500 ft)(2.58)(62.4 lb per cu ft)/144 psf], the safety factor will have to be:

$$F = (1 - 0.75)\left(\frac{15{,}000 \text{ psi}}{559 \text{ psi}}\right) = 6.7$$

Solution 4.7

The compressive strength of the cube is:

$$\sigma_c = \frac{8{,}000 \text{ lb}}{4 \text{ sq in.}} = 2{,}000 \text{ psi}$$

$$K = \sigma_c(D)^{0.5} = 2{,}000(2)^{0.5} = 2{,}828$$

$$\sigma_p = \sigma_v \left(\frac{1}{1-R}\right) = 500(1.1)\frac{1}{0.5} = 1{,}100 \text{ psi}$$

$$\sigma_p = \frac{K(W)^{0.5}}{H} = \frac{2{,}228(W)^{0.5}}{(5)(12)}$$

$$47.13(W)^{0.5} = 1{,}100$$

$$(W)^{0.5} = 23.34$$

$$W = 544.74 \text{ in.} = 45.4 \text{ ft}$$

Solution 4.8

First, calculate the average pillar stress:

average pillar stress = (premining stress)(tributary area/pillar area)

$$= 750(1.1)[(66^2)/(50^2)]$$

$$= 1,437 \text{ psi}$$

Next, calculate the pillar strength:

$$\sigma_1 = \frac{K}{(36)^{0.5}} = \frac{4,000}{(36)^{0.5}} = 667 \text{ psi}$$

$$\sigma_p = \sigma_1[0.64 + 0.36\left(\frac{W}{h}\right)] = (667)[0.64 + 0.36\left(\frac{50}{6}\right)] = 2,428 \text{ psi}$$

Finally, calculate the stability factor (SF):

$$SF = \frac{\text{pillar strength}}{\text{pillar stress}} = \frac{2,428}{1,437} = 1.69$$

The stability factor is less than 2.0, and therefore a larger pillar size should be investigated.

Solution 4.9

Federal law dictates that the maximum entry width in coal mines where only one means of roof support is used is 20 ft.

$$R = \frac{A_t - A_p}{A_t} = 0.5 = \frac{(W+20)^2 - W^2}{(W-20)^2}$$

Solving, W = 48.28 ft. Is a 48.28-ft pillar sufficiently large? First, determine the pillar stress, taking into account the recovery:

$$\text{Pillar stress: } \sigma_v \left(\frac{1}{1-R}\right) = (1.1)(650)\left(\frac{1}{1-0.5}\right) = 1,430 \text{ psi}$$

Next, determine the pillar strength (remember that W and H are in inches):

$$K = S_c(D)^{0.5} = 2,826(4)^{0.5} = 5,652$$

$$\sigma_p = \frac{K(W)^{0.5}}{H} = \frac{(5,652)(579)^{0.5}}{72} = 1,889 \text{ psi}$$

Because, the safety factor is 1.32 (1,889/1,430), and the recommended safety factor for use with the Holland–Gaddy equation is between 1.8 and 2.2, it appears that the pillar is undersized.

Solution 4.10

Premining stress = (600 ft)[(160 lb per cu ft)/(144 sq in. per sq ft)] = 667 psi

The total load on the pillar is equal to the premining stress multiplied by the tributary area:

$$(667 \text{ psi})(L + 20)^2$$

Multiplying the above-mentioned value by the safety factor and dividing by the cross-sectional area of the pillar equals the strength in compression, from which L can be derived:

$$\frac{(667 \text{ psi})(L + 20)^2 (2.5)}{L^2} = 9{,}000 \text{ psi}$$

Thus, L = 15 ft

The pillar's W/H ratio is 1.25(15/12), which differs from the specimen W/H ratio. Therefore, use Eq. 4.6 to determine C_1 for the specimen.

$$\sigma_p = C_1 [0.778 + 0.222\left(\frac{W}{H}\right)]$$

$$9{,}000 = C_1 [0.778 + 0.222(0.5)]$$

$$C_1 = 10{,}124 \text{ psi}$$

For the pillar:

$$\sigma_p = 10{,}124 \, [0.778 + 0.222(180/144\,)]$$

$$= 10{,}686 \text{ psi}$$

L should, therefore, be recalculated:

$$\frac{(667 \text{ psi})(L + 20)^2 (2.5)}{L^2} = 10{,}686$$

$$\frac{(L + 20)^2}{L^2} = 6.4$$

$$L = 13 \text{ ft}$$

Therefore, under these mining conditions, the pillar should be at least 13 ft × 13 ft. The initial extraction ratio is:

$$R = 1 - \frac{FS_v}{C_p}$$

$$= 1 - \frac{(2.5)(667)}{10{,}686}$$

$$= 0.84$$

Solution 4.11

According to 30 CFR 75.222 (b), the maximum distance between roof bolts in a pattern is 5 ft × 5 ft. If a 20-ft-wide entry is advanced 20 ft, as shown in the following plan view, the minimum number of bolts required is 12.

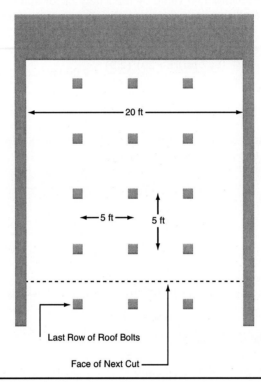

Illustrates Solution to Problem 4.11.

The allowable load per bolt incorporating #6 rebar grade 60 is:

$$A_L = \text{(cross-sectional area of the bolt)}(60{,}000 \text{ psi})(\text{yield factor})$$

$$= \left(\frac{0.75}{2}\right)^2 \pi (60{,}000)(0.8)$$

$$= 21{,}206 \text{ lb}$$

The rock load of the cut is:

$$L = (20 \text{ ft})(20 \text{ ft})(5 \text{ ft})(160 \text{ lb per cu ft})$$

$$= 320{,}000 \text{ lb}$$

The rock load per bolt is:

$$\frac{320{,}000}{12} = 26{,}667 \text{ lb}$$

Therefore, since 26,667 > 21,206, the proposed bolting pattern is inadequate. It is recommended that the pattern be modified with a closer spacing of bolts and recalculated.

Solution 4.12

Using Eq. 4.13,

$$L = \left(\frac{4t^2 \sigma_e}{3W}\right)^{0.5}$$

where:

$t = 2$ ft

$\sigma_e = \left(\frac{625 \text{ psi}}{8}\right)(144 \text{ sq in. per sq ft}) = 11{,}250$ psf

$W = (2.4)(62.4 \text{ lb per sq ft}) = 150$ lb per cu ft

$$L = \left(\frac{(4)(2 \text{ ft})^2(11{,}250 \text{ psf})}{3(150 \text{ lb per sq ft})}\right)^{0.5}$$

$= (400 \text{ sq ft})^{0.5} = 20$ ft

Solution 4.13

When the entries (A) and crosscuts (B) are driven 18 ft wide, the diagonal intersection span (I_S) is $I_S^2 = A^2 + B^2 = 18^2 + 18^2 = 324 + 324 = 648$; thus, $I_S = 25.5$ ft.

Substituting into Eq. 4.16 results in the following bolt length:

$$L_B = 0.12 I_s \log_{10} H \left(\frac{100 - \text{CMRR}}{100}\right)$$

$$= (0.12)(25.5)\log_{10}(600)\left(\frac{100 - 45}{100}\right)$$

$= (0.12)(25.5)(2.78)(0.55)$

$= 4.68$ ft

Evaluate a 5-ft bolt, but consider a 6-ft bolt because of the bed of sandstone just above the laminated shales.

Solution 4.14

Using Eq. 4.14:

5 rows, therefore, $n_1 = 5$

3 bolts per row, therefore, $n_2 = 3$

$$P = \frac{wtBL}{(n_1 + 1)(n_2 + 1)}$$

where w is 160 lb per cu ft, t is 4 ft, B is 16 ft, and L is 20 ft.

Therefore,

$$P = \frac{(160)(4)(16)(20)}{(6)(4)} = 8{,}533 \text{ lb per bolt}$$

Because grade-40 steel is rated at 40,000 psi, multiplying the load per bolt by the safety factor and dividing by 40,000 psi yields the cross-sectional area of the bolt:

$$\frac{(8{,}533 \text{ lb})(2)}{40{,}000 \text{ psi}} = A = 0.427 \text{ sq in.} = \pi r^2; \; r = 0.369$$

Thus, the bolt's diameter is $2r = 2(0.369) = 0.738$ in. A 0.75-in. bolt is recommended.

Solution 4.15

This problem illustrates convergence in mine openings.

Sag of the immediate roof: Because the base of point A is considered the stable point, the closure between points A and B represents the sag of the immediate roof:

$$50.3 \text{ in.} - 46.1 \text{ in.} = 4.2 \text{ in.}$$

Heave of the immediate floor: Because the sag of the immediate roof is 4.2 in. and the total closure between points D and C can be calculated as follows:

$$52.3 \text{ in.} - 46.3 \text{ in.} = 6.0 \text{ in.}$$

the difference between the two values can be attributed to floor heave:

$$6.0 \text{ in.} - 4.2 \text{ in.} = 1.8 \text{ in.}$$

Bed separation between the immediate and main roofs: Because the difference between the closure of points C and D (6.0 in.) and points E and F (2.6 in.) represents the separation of the immediate roof from the main roof, the value can be calculated as follows:

$$6.0 \text{ in.} - 2.6 \text{ in.} = 3.4 \text{ in.}$$

Solution 4.16

The weight of the rock column to be supported is:

$$(72 \text{ ft})(14 \text{ ft})(5 \text{ ft})(0.08 \text{ tons per cu ft}) = 403.2 \text{ tons}$$

At a support density of 10.0 tons per sq ft, the rating for the shield is:

$$(10 \text{ tons per sq ft})(11 \text{ ft})(5 \text{ ft}) = 550 \text{ tons}$$

Because 550 tons > 403.2 tons, the shields should be adequate.

Solution 4.17

The width-to-depth ratio is 0.8(800/1000). By utilizing the curve shown below, the S/H ratio that corresponds to the above-mentioned value is 0.75.

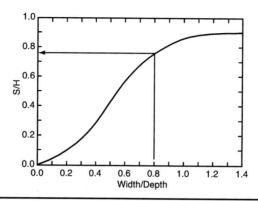

Illustrates Solution to Problem 4.17

If the height of the coal seam is 6 ft, the maximum anticipated subsidence would be 4.5 ft [(6)(0.75)].

Solution 4.18

If the maximum allowable subsidence of the surface is 2 ft and the seam is 5-ft thick, the S/H ratio is 0.4. Using the NCB's curve shown below, the corresponding W/D ratio is 0.5. At a depth of 1,200 ft, therefore, the maximum allowable opening is 600 ft. Since this is considered an undesirably short longwall face length by today's standards, additional precautions must be taken to protect the pipeline.

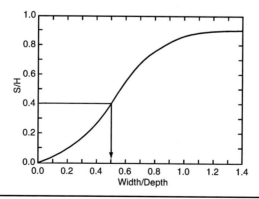

Illustrates Solution to Problem 4.18

Solution 4.19

The surface slopes less than 5%; no downslope-stability requirements are necessary.

The support-area dimensions are calculated as follows:

	Building Dimension	+	(2) (offset distance)	+	2 (D) (tan 15°)
Shorter side:	30	+	(2) (15)	+	2 (600) (0.27)
Longer side:	100	+	(2) (15)	+	2 (600) (0.27)

Thus, the support area is 384 ft × 454 ft.

Solution 4.20

Since the support area is 350 ft × 450 ft, the in-place tonnage is:

$$\frac{(6 \text{ ft})(350 \text{ ft})(450 \text{ ft})(80 \text{ lb per cu ft})}{(2{,}000 \text{ lb per ton})} = 37{,}800 \text{ tons}$$

The additional tonnage that can be extracted if the building is purchased is (37,800 tons)(0.30) = 11,340 tons. At a profit of $3.50 per ton, the additional coal is worth $39,690. Under these circumstances, the mine should consider purchasing the structure.

Solution 4.21

The critical width is defined as the opening width beyond which no additional vertical surface subsidence occurs. Expressed mathematically:

Critical width = 2 [tan(angle of draw)](depth of cover)

= 2(tan 25°)(1,000 ft)

= 933 ft

If the opening width is 933 ft and the depth of cover is 1,000 ft, the W/D ratio is 0.933. This implies that no additional vertical subsidence occurs above this value, but, referring to the NCB's subsidence development curve, the curve does not level off until the W/D ratio exceeds approximately 1.2. Thus, it appears that the angle of draw in the areas analyzed by the NCB was somewhat greater than 25°.

Solution 4.22

1. Calculate the width-to-depth ratio.

$$\frac{W}{D} = \left(\frac{1{,}000 \text{ ft}}{1{,}430 \text{ ft}}\right) = 0.70$$

2. Determine the subsidence factor (S/H) from the NCB graph showing the relationship of subsidence to the width-to-depth ratio. From the graph shown below, the value for S/H is approximately 0.675.

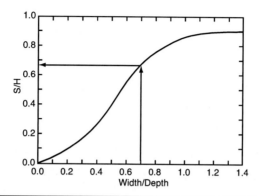

Illustrates Solution to Problem 4.22

3. Determine the maximum anticipated subsidence.

$$\text{Subsidence}_{max} = (\text{seam thickness})(\text{subsidence factor})$$

$$= (6 \text{ ft})(0.675) = 4.05 \text{ ft}$$

4. Using the values from Table 4.2, which indicate the relationship between W/D and distances from the panel center in terms of depth (d/D), tabulate the transverse subsidence profile.

Distance from Panel Center Line		Subsidence	
d/D (from chart)	d (calculated)	s/S (from chart)	s (calculated)
1.05	(1,430)(1.05) = 1,502	0.00	0.00
0.49	701	0.10	0.41
0.39	558	0.20	0.81
0.33	472	0.30	1.22
0.29	415	0.40	1.62
0.25	358	0.50	2.03
0.21	300	0.60	2.43
0.18	257	0.70	2.84
0.14	200	0.80	3.24
0.10	143	0.90	3.65
0.00	0	1.00	4.05

5. Using the values from Table 4.3, which indicate the relationship for various strain values in a subsidence profile and the graph for predicting maximum slope and strain for various width-to-depth ratios of the panel, determine the transverse strain profile, the maximum extension, the maximum compression, and the profile data.

$$\text{Transverse strain profile} = (S/D) = (4.05/1{,}430) = 0.0028$$

From the graph, the multiplier for extension is 0.69, and the multiplier for compression is 0.84.

$$\text{Maximum extension} = (0.69)(0.0028) = 0.0019 \text{ ft per ft}$$

$$\text{Maximum compression} = (0.84)(0.0028) = 0.0024 \text{ ft per ft}$$

Distance (from panel center line)		Strains			
		Extension (+E)		Compression (−E)	
d/D (from chart)	d (calculated)	(chart)	(calculated) ft per 1,000 ft	(chart)	(calculated) ft per 1,000 ft
1.05	(1,430)	(1.05)	= 1,502	0.00	0.00
0.65	930	0.20	0.38		
0.54	772	0.40	0.76		
0.48	686	0.60	1.14		
0.44	629	0.80	1.52		
0.37	529	1.00	1.90		
0.33	472	0.80	1.52		
0.25	358	0.00	0.00	0.00	0.00
0.23	329			0.20	0.48
0.20	286			0.40	0.96
0.17	243			0.60	1.44
0.14	200			0.80	1.92
0.08	114			1.00	2.40
0.00	0			0.80	1.92

At 500 ft from the panel center line (the edge of the opening), the maximum strain across the structure is between 1.52 ft per 1,000 ft (extension at 472 ft) and 1.90 ft per 1,000 ft (extension at 529 ft). We shall use 1.7 ft per 1,000 ft or 0.0017 for the damage assessment.

Using the damage classification chart shown above, a strain of approximately 0.0017 would be in the slight range for a structure 30 m (≈ 100 ft) long.

Solution 4.23

As calculated in problem 4.22, the various subsidence values associated with distances from the panel center line are indicated as shown.

Plotting these pairs of points on the diagram results in the following:

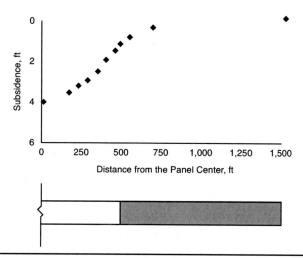

Illustrates Solution to Problem 4.23

Distance from Panel Center Line (ft)	0	143	200	257	300	358	415	472	558	701	1,502
Calculated Subsidence (ft)	4.05	3.65	3.24	2.84	2.43	2.03	1.62	1.22	0.81	0.41	0

Solution 4.24

The hyperbolic tangent function is:

$$S(x) = \frac{0.5}{S_{max}/\left(1 - \tanh\frac{cx}{b}\right)}$$

where:
1. $S(x)$ is the surface subsidence for various distances from the rib (x) of a given panel. This will be calculated.
2. S_{max} is the maximum subsidence observed over a given panel. This will be assumed to be 4.05 ft.

3. b is the distance from the inflexion point, or point of maximum slope, to the location where the curve flattens out at maximum subsidence. Since the point of inflexion is assumed to be directly above the edge of the opening in the above equation, b = 500 ft.
4. c is a constant representing a characteristic of the type of subsidence that occurred, such as subcritical, critical, or supercritical. Since this will obviously be in the supercritical range, its corresponding value (2.0) will be used.

The values used to calculate the subsidence at the panel center are as follows:

S_{max} = 4.05 ft

c = 2.0

x = −500 ft (since the center is 500 ft to the left of the rib)

b = 500 ft

$$S(x) = (0.5)((S_{max})(1 - \tanh \frac{cx}{b})$$

$$= (0.5)((4.05)(1 - \tanh ((2.0)\left(\frac{-500}{500}\right)))$$

$$= (0.5)((4.05)(1 - (-0.96))$$

$$= (0.5)((4.05)(1.96))$$

$$= 4.0 \text{ ft}$$

The subsidence values for points located from the panel center to the right rib and beyond, at 100-ft intervals, are as follows:

S(panel center) = (0.5) ((4.05) (1 − tanh ((2.0)(−500)/(500))) = 4.0

S(400 ft left of rib line) = (0.5) ((4.05) (1 − tanh ((2.0)(−400)/(500))) = 3.9

S(300 ft left of rib line) = (0.5) ((4.05) (1 − tanh ((2.0)(−300)/(500))) = 3.7

S(200 ft left of rib line) = (0.5) ((4.05) (1 − tanh ((2.0)(−200)/(500))) = 3.4

S(100 ft left of rib line) = (0.5) ((4.05) (1 − tanh ((2.0)(−100)/(500))) = 2.8

S(right rib line) = (0.5) ((4.05) (1 − tanh ((2.0)(0)/(500))) = 2.0

S(100 ft right of rib line) = (0.5) ((4.05) (1 − tanh ((2.0)(100)/(500))) = 1.3

S(200 ft right of rib line) = (0.5) ((4.05) (1 − tanh ((2.0)(200)/(500))) = 0.7

S(300 ft right of rib line) = (0.5) ((4.05) (1 − tanh ((2.0)(300)/(500))) = 0.3

S(400 ft right of rib line) = (0.5) ((4.05) (1 − tanh ((2.0)(400)/(500))) = 0.2

S(500 ft right of rib line) = (0.5) ((4.05) (1 − tanh ((2.0)(500)/(500))) = 0.1

Plotting these points on the same diagram used for the NCB-derived profile results in the following:

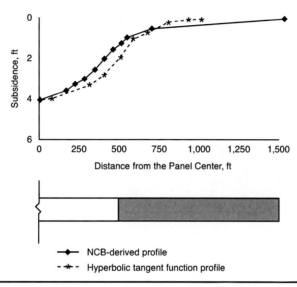

Illustrates Solution to Problem 4.24

Solution 4.25

$$F = \frac{cA + (W\cos\psi_p - U - V\sin\psi_p)\tan\phi}{W\sin\psi_p + V\cos\psi_p}$$

This equation can be reduced to the following form if the influence of water pressure is ignored:

$$F = \frac{cA + (W\cos\psi_p)\tan\phi}{W\sin\psi_p}$$

The length of the slip surface, A, can be determined, through the geometry of the deposit, to be 90 ft:

The weight of the block can be calculated based on a face length of 1 ft:

Total volume = (78)(100)(1) =		7,800 cu ft
− Triangle A = (0.5)(58)(100)(1) =		−2,900 cu ft
− Triangle B = (0.5)(45)(78)(1) =		<u>−1,755 cu ft</u>
		3,145 cu ft

MINING ENGINEERING ANALYSIS

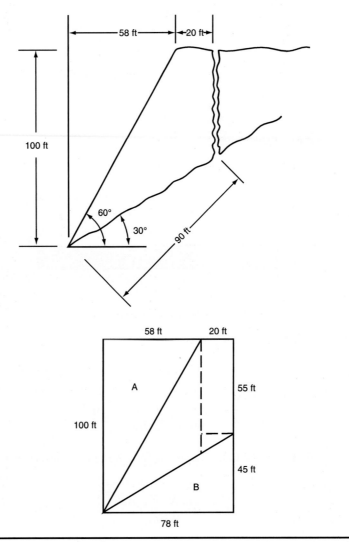

Illustrates Solution to Problem 4.25

Therefore,

$$W = (3{,}145 \text{ cu ft})(160 \text{ lb per cu ft})$$
$$= 503{,}200 \text{ lb}$$
$$\psi_p = 30°$$
$$\varphi = 25°$$
$$c = 1{,}000 \text{ psf}$$

$$F = \frac{cA + (W\cos\psi_p)\tan\varphi}{W\sin\psi_p}$$

$$F = (1{,}000 \text{ psf})(90 \text{ sq ft}) + \frac{(503{,}200 \text{ lb})(\cos 30°)\tan 25°}{(503{,}200 \text{ lb})(\cos 30°)}$$

$$= \frac{90{,}000 \text{ lb} + 203{,}209 \text{ lb}}{251{,}600 \text{ lb}} = 1.17$$

Solution 4.26

The water would stand a vertical distance of 35 ft in the crack:

$$100 \text{ ft} - [(\sin 30°)(90)] - 20 \text{ ft} = 35 \text{ ft}$$

$$V = (0.5)(\gamma_w)(Z_w^2)(\sin\psi)$$
$$= (0.5)(62.4 \text{ lb per cu ft})(35 \text{ ft})^2(1.0 \text{ ft})$$
$$= 38{,}220 \text{ lb}$$

$$U = (0.5)(\gamma_w)(Z_w)(H - Z)(\operatorname{cosec}\psi_p)$$
$$= (0.5)(62.4 \text{ lb per cu ft})(35 \text{ ft})(100 \text{ ft} - 35 \text{ ft})(2.0)(1.0 \text{ ft})$$
$$= 141{,}960 \text{ lb}$$

Now, direct substitution into Eq. 4.20 is possible:

$$F = \frac{90{,}000 \text{ lb} + [435{,}611 \text{ lb} - 141{,}960 \text{ lb} - 38{,}220 \text{ lb}(0.5)](0.466)}{251{,}600 + (38{,}220)(0.866)}$$

$$= 0.77$$

Thus, the presence of water in the tension crack creates a hazardous situation (F < 1.0).

CHAPTER 5

Pumping and Drainage

It is extremely important to control nuisance water in a mine; left uncontrolled, nuisance water can have a severe effect on haulage, ventilation, production, and the health and safety of miners. To control water in mines, a four-step process is followed: (1) prevention, (2) collection, (3) transportation, and (4) treatment. This chapter discusses the collection and transportation of mine water, because prevention is usually dictated by premining conditions and initial planning, while treatment is variable and highly site specific.

A mine drainage system is nothing more than a collection of gathering points (sumps), machines that impart energy to the fluid (pumps), transportation ducts (pipes), and control devices (fittings, valves, etc.). A pump does not pull the fluid up the suction pipe; instead, atmospheric or external pressure pushes the liquid up the suction pipe to the pump.

All references in this chapter will be to centrifugal pumps, due to their popularity for mine drainage.

PUMP CHARACTERISTIC CURVES

Pump performance is plotted on a graph as an indication of its characteristic. To develop the data points, a test facility similar to the one shown in Figure 5.1 is established. The calibrated Venturi meter is used to measure quantity, the Bourdon gage measures pressure, and the wattmeter measures power input. Various discharge pressures are obtained by throttling with a gate valve in the discharge pipe. Quantity and power are measured at each of these pressure levels, giving the necessary data for plotting the characteristic curves.

Figure 5.2 shows a typical characteristic curve for a centrifugal pump. Notice that pressure is measured as so many feet of vertical water column above the outlet level, or head. Knowing the quantity (in gallons per minute), head (in feet), and power input (as brake horsepower), the pump's efficiency can be plotted based on the following relationship:

$$E = \frac{QH(8.33 \text{ lb of water per gal})}{(33{,}000 \text{ ft-lb per min})(\text{brake horsepower})} \quad \text{(EQ 5.1)}$$

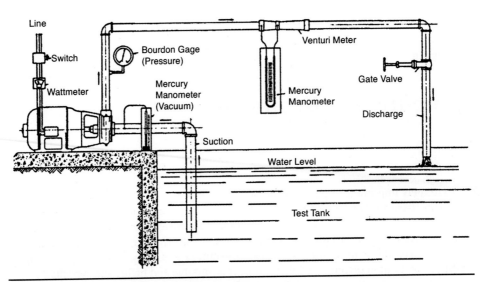

FIGURE 5.1 Centrifugal pump testing arrangement
Source: Myers Co. 1981.

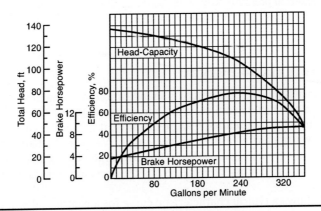

FIGURE 5.2 Typical characteristic curves for a centrifugal pump

PIPE CHARACTERISTIC CURVES

The total head against which a pump operates is based on the following equation:

$$H = h_s + h_f + h_v + h_{sh} \qquad \text{(EQ 5.2)}$$

where h_s is the vertical distance, in feet, from the suction liquid level to the discharge liquid level (total static head); h_f is the equivalent head, expressed as feet of liquid, required to overcome the friction caused by the flow through the pipe (friction head); h_v is the head, in feet, required to create velocity of flow (velocity head); and h_{sh} is the head, in feet, required to overcome the shock losses due to changes of water flow produced by fittings.

TABLE 5.1 Equivalent number of feet of straight pipe for different fittings

	Size of Fittings, in.						
	½	1	2	3	4	5	6
90° ell	1.5	2.7	5.5	8.0	11.0	14.0	16.0
45° ell	0.8	1.3	2.5	3.8	5.0	6.3	7.5
Long ell	1.0	1.7	3.5	5.2	7.0	9.0	11.0
Tee straight	1.0	1.7	3.5	5.2	7.0	9.0	11.0
Tee side	3.3	5.7	12.0	17.0	22.0	27.0	33.0
Gate valve, open	0.4	0.6	1.2	1.7	2.3	2.9	3.5
Check valve	4.0	8.0	19.0	32.0	43.0	58.0	100.0

Note that, in most cases, this value is negligible and is often ignored. Table 5.1 shows the equivalent number of feet of straight pipe for different fittings.

Frictional head is usually expressed using the following equation based on the number of 100-ft lengths of pipe in the system:

$$h_f = 0.2083(100/C)^{1.85} \left(\frac{q^{1.85}}{d^{4.8655}} \right) \qquad \text{(EQ 5.3)}$$

where C is a constant, usually 100, accounting for surface roughness; q is the flow, in gallons per minute; and d is the inside diameter of the pipe, in inches. The frictional-head values are often tabulated as shown in Table 5.2.

Velocity head, though often ignored, can be calculated using the following formula:

$$h_v = \frac{v^2}{2g} \qquad \text{(EQ 5.4)}$$

where v is the velocity of the liquid, in feet per second, and g is the acceleration due to gravity, in feet per second per second (32.2 fpsps).

It is generally considered good practice to keep the velocity in the pipeline below 8 fps on the suction side and below 15 fps on the discharge side of the pump. This prevents the friction losses from becoming excessive. Figure 5.3 is a schematic showing the various head components.

By selecting various quantities, the total head can be represented by a curve called the pipe characteristic curve. When plotted on the graph containing the pump characteristic curve, a point of intersection, known as the operating point, is determined.

PUMP FLEXIBILITY

A pump is often inappropriate for a given set of conditions (i.e., no point of intersection with the pipe curve, the operating point is at a lower quantity than the inflow rate, etc.). In these instances, the pump can be altered by staging or by a speed change.

Two or more pumps may be placed in series or a single pump may be equipped with more than one impeller on the same shaft (multistage operation) in order to operate at

TABLE 5.2 Friction loss, in feet, for old pipe (C = 100)

Flow, gpm	Pipe Size, in.							Flow, gpm
	½	1	2	3	4	5	6	
2	7.4							2
4	27.0	2.14						4
6	57.0	4.55	0.20					6
8	98.0	7.80	0.33					8
10	147.0	11.70	0.50	0.07				10
15		25.00	1.08	0.15				15
20		42.00	1.82	0.25	0.06			20
30		89.00	3.84	0.54	0.13	0.04		30
40		152.00	6.60	0.91	0.22	0.08		40
50			9.90	1.38	0.34	0.11	0.05	50
60			13.90	1.94	0.47	0.16	0.07	60
80			23.70	3.30	0.81	0.27	0.11	80
100			35.80	4.98	1.23	0.41	0.17	100
150			76.00	10.60	2.61	0.88	0.37	150
200			129.00	18.00	4.43	1.50	0.62	200
250				27.20	6.76	2.27	0.94	250
300				38.00	9.30	3.17	1.30	300
350				50.60	12.50	4.22	1.74	350
400				64.70	16.00	5.40	2.22	400
450				80.50	19.90	6.65	2.76	450
500				97.80	24.10	8.15	3.36	500
600				137.00	33.80	11.70	4.70	600
700					45.00	15.20	6.25	700
800					57.60	19.40	8.00	800
900					71.60	24.20	9.95	900
1,000						29.40	12.10	1,000
1,200						41.10	16.90	1,200

higher heads. In these cases, the heads and horsepowers are additive at equivalent quantities, as shown in Figure 5.4.

When the speed of a centrifugal pump is changed, the operation of the pump is changed in accordance with the following three fundamental laws:

- Q varies directly as the speed.
- H varies as the square of the speed.
- Brake horsepower varies as the cube of the speed.

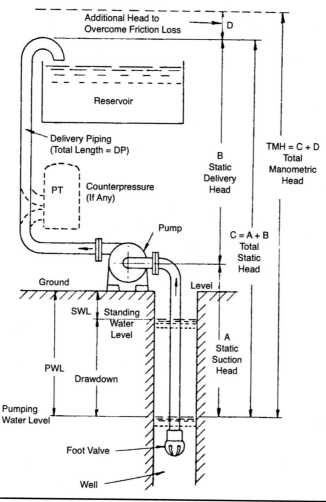

FIGURE 5.3 Components of head
Source: Hayward Tyler, Inc. 1985.

There is a limit to the amount that the speed of a pump can be increased, however. To avoid pump damage, the limit recommended by the manufacturer should always be observed.

PUMPING APPLICATIONS

Pumps are basically used in two ways in mine drainage situations: (1) station duty and (2) dewatering.

A typical station-duty application consists of a pump with a short suction line (less than 20 ft) that removes water from a large sump and discharges it against a constant

126 | MINING ENGINEERING ANALYSIS

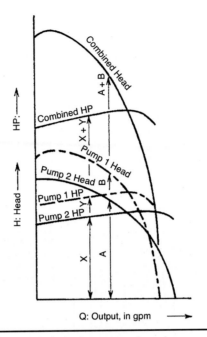

FIGURE 5.4 Combined characteristic for two pumps in series

static head. The discussion, up to now, has focused on determining the operating points for station-duty pumps.

A dewatering system is much more complex than a station-duty system. An example of a dewatering application is the removal of water from a flooded air shaft or mine with a submersible pump. Assuming that the pumping rate always exceeds the inflow rate, the pump must operate over a wide and varying range of static heads, i.e., from a minimum static head when pumping commences, to a maximum static head when pumping terminates. Furthermore, during pumping, an inflow can keep recharging the standing-water capacity to some degree. To solve problems of this nature, the "method of horizontal slices" is recommended. Basically, the method consists of the following steps:

1. Secure the curves of a multistage pump to be analyzed.
2. Plot pipe curves at the following locations:
   a. The elevation of the standing water level, which represents the initial conditions.
   b. The elevation when the mine or shaft is dry, which represents the final conditions.
   c. Each elevation between the two aforementioned extremes (2a and 2b) where the geometry of the water column's plan view changes.
   d. Any other elevation in between the two extremes if increased accuracy is desired.

3. Determine if the curve representing the final conditions intersects the pumping curve at a value greater than the inflow rate.
   a. If yes, move to step 4.
   b. If no, increase the pumping curve in step 1 by one stage and reevaluate the question posed in step 3.
4. Determine the quantity that corresponds to each of the pipe/pump curve intersections.
5. Evaluate steps 5a, 5b, and 5c, in order, for as many instances as there are intervals.
   a. Determine the average quantity (Q_{avg}) for each interval starting at water level and going downward until the mine or shaft is pumped dry.
   b. Multiply each interval's cross-sectional area (in square feet) by its corresponding height to get the cubic feet of water in each interval. Multiplying by 7.48 converts the cubic feet into gallons of water.
   c. To calculate the time to pump the water from each interval, divide the total gallons in each interval by the difference between the interval's average pumping quantity and the inflow rate. Be careful that the inflow rate is given in gallons per minute.
6. Sum up all the values for the pumping times in each interval to come up with the time (in minutes) it would take to pump all the water.
7. Sum up the quantities in all of the intervals, in gallons. This summation represents the standing water (i.e., does not include the inflow).
8. To calculate the representative gallons per minute, substitute into the following equation:

$$\text{gpm} = \frac{(\text{standing water}) + (\text{total time})(\text{inflow rate})}{(\text{total time})} \qquad \text{(EQ 5.5)}$$

REFERENCES

F.E. Myers Co. 1981. *Centrifugal Pump Manual.* Ashland, OH: F.E. Myers Company.
Goulds Pumps, Inc. 1982. *GPM (Goulds Pump Manual).* Seneca Falls, NY: Goulds Pumps, Inc.
Hayward Tyler, Inc. 1985. *Total Pumping Head and Pump Selection, Technical Notes No. 1.* Norwalk, CT: Hayward Tyler, Inc.
Westaway, C.R., and A.W. Loomis. 1979. *Cameron Hydraulic Data.* Woodcliff Lake, NJ: Ingersoll-Rand Company.

PROBLEMS

Problem 5.1

If atmospheric pressure "pushes" mine water up a suction line due to the vacuum created by a pump, is there a limitation to the maximum length of suction line? If so, what is the length, in feet?

Problem 5.2

What is the brake horsepower required to pump 150 gpm against a total dynamic head of 370 ft if the pump operates at 70% efficiency?

Problem 5.3

If a 1,500-ft-long 5-in. pipe carries 150 gpm of mine water, what is the head loss due to friction?

Problem 5.4

If the velocity of water flowing through a pipe is 900 fpm, what is the velocity head?

Problem 5.5

What is the equivalent number of feet of straight pipe for two 5-in. check valves?

Problem 5.6

Given the station-duty pumping system shown in the following figure, and noting that there is one 90° ell on the suction line and three 90° ells on the discharge line, determine the total dynamic head that the pump must be able to overcome. Assume that the water level in the sump remains constant.

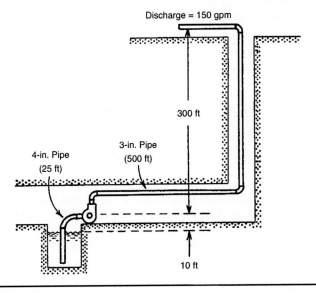

Illustrates Problem 5.6

Problem 5.7

If the efficiency of a pump is determined to be 78% and the efficiency of the motor that drives the pump (motor drive efficiency) is 84%, what is the overall efficiency of the pumping system?

Problem 5.8

A pump, operating at 200 gpm and 168 ft of head, has the pumping and motor drive efficiencies noted in problem 5.7. Compute the power consumption, in kilowatts, for this situation.

Problem 5.9

The characteristic curves shown in Figure 5.2 are based on a pump that operates at 3,450 rpm. What would the curves look like in the interval between 280 and 360 gpm if the pump's speed were increased to 3,900 rpm? What is the speed ratio?

Problem 5.10

The centrifugal pump, whose characteristic curves are shown in Figure 5.2, operates against a static head of 80 ft in connection with a 1,000-ft 5-in. discharge line. Neglecting shock losses, velocity head, and friction head in the suction line, determine the operating point of the pump.

Problem 5.11

If the inflow rate into the sump for problem 5.10 were in excess of 250 gpm, it should be obvious that the pumping system is insufficient. Suppose, for example, that the inflow rate is 320 gpm and the pump's speed is to be adjusted to accommodate the new situation. What must be the new speed ratio to pump this amount?

Problem 5.12

A 22-ft-diameter air shaft was sunk 300 ft to the bottom of a coal seam. It was subsequently abandoned and allowed to fill with water to within 25 ft of the collar at an inflow rate of 400 gpm. The company owns a multistage centrifugal pump and a considerable amount of 5-in. pipe. The single-stage characteristic of the multistage submersible pump is shown.

Determine how long it will take to dewater the shaft with this combination using the method of horizontal slices. Pass slices through the shaft as established in the criteria in this chapter, as well as at static heads of 100 ft and 200 ft. Neglect shock losses and velocity head. What is the minimum number of stages required? If the motor drive efficiency is 80% and power costs are $0.08 per kW-hr, what will be the operating cost to dewater the shaft?

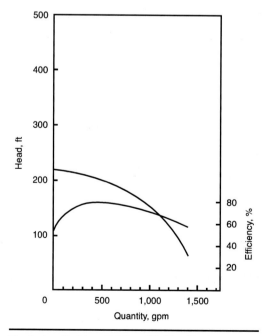

Illustrates Problem 5.12 and 5.13—Single-stage characteristic of the multistage submersible pump

Problem 5.13

The flooded limestone mine shown in the following figure is to be dewatered by a multistage submersible pump.

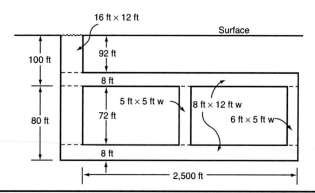

Illustrates Problem 5.13—Flooded limestone mine

The single-stage characteristic of the multistage submersible pump is shown.

An overflow of 380 gpm is measured at the shaft collar (inflow rate). If power costs $0.08 per kW-hr, the mine has a sufficient quantity of 5-in. pipe, and the motor drive efficiency is 82%, determine the minimum number of stages, the time it would take to pump out the mine, and the total operating costs using the method of horizontal slices. Use the minimum number of slices recommended by the criteria established in this chapter and recall that there are 7.48 gal of water per cubic foot.

Problem 5.14

A pumping system has an operating point of 200 gpm at 164 ft and an efficiency of 78%. If the inflow rate is 105 gpm, determine the daily pumping hours required.

Problem 5.15

To take advantage of reduced power costs passed on to the mine by the utility, management would like to perform all pumping during the 8-hr off-peak period. What would the pumping rate for the installation in problem 5.14 need to be in order to handle the daily inflow in 8 hours?

Problem 5.16

After a speed change, the operating point for problem 5.15 was determined to be 315 gpm at 178 ft and an efficiency of 63%. Was it economically advantageous to increase the pump speed, over the situation in problem 5.14, if the 16-hr on-peak cost is $0.10, the 8-hr off-peak cost is $0.08, and the motor drive efficiency is 80%?

Problem 5.17

A main sump is being planned for the bottom area of a coal mine. It is projected to be 20 ft wide and 220 ft long and it will be installed as rapidly as possible after the slope reaches the coal seam. While the bottom area is under development, a single-action piston pump will be connected to the sump, to be permanently replaced at a later date. The pump's piston is 6 in. in diameter, and the length of its stroke is 14 in.; normally it operates at 30 strokes per minute. The sump will be designed for a depth of 6 ft, but it is felt that the standing water level should be kept as close to 3 ft as possible. If, after a heavy inflow, the sump has a standing water level of 4.5 ft and the current inflow has reduced to a 30-gpm level, determine how many days the pump will have to operate (if it normally will run 16 hr per day) to lower the water level in the sump to the designed standing water level.

PROBLEM SOLUTIONS

Solution 5.1

At sea level, atmospheric pressure is equal to 14.7 psi. If a perfect vacuum were to be created in a suction line, atmospheric pressure could push a 1-in. column of water to a height of

$$\text{pressure} = \text{weight of water column}$$

$$14.7 \text{ lb per sq in.} = (1 \text{ in.})(1 \text{ in.})(H \text{ in.})\left(\frac{62.4 \text{ lb}}{1{,}728 \text{ cu in.}}\right)$$

$$\frac{(14.7 \times 1{,}728)}{62.4} = H \text{ in.}$$

$$H = 407 \text{ in.} = 33.9 \text{ ft}$$

This shows that the maximum total suction lift, when pumping water at sea level, is approximately 34 ft. This figure is theoretical and can never be obtained in practice. That is why good mining practice dictates that the total dynamic suction lift should never exceed a value of approximately 20 ft for trouble-free operation. For a more accurate treatment of this situation, consult a pump manufacturer's catalog with reference to a particular pump's net positive suction head (NPSH).

Solution 5.2

$$HP_B = \frac{QH(8.33)}{(33{,}000)(E)}$$

$$= \frac{(150)(370)(8.33)}{(33{,}000)(0.7)}$$

$$= 20.01 \text{ hp}$$

Solution 5.3

$$h_f = (0.2083)\left(\frac{100}{C}\right)^{1.85}\left(\frac{q^{1.85}}{d^{4.8655}}\right)\left(\frac{1{,}500}{100}\right)$$

132 | MINING ENGINEERING ANALYSIS

$$= (0.2083)\left(\frac{100}{100}\right)^{1.85}\left(\frac{150^{1.85}}{5^{4.8655}}\right)(15)$$

$$= 0.2083(1)(4.22)(15)$$

$$= 13.19 \text{ ft}$$

Also, using Table 5.2:

$$h_f = 0.88\left(\frac{1,500}{100}\right) = 13.2 \text{ ft}$$

Solution 5.4

$$h_v = \frac{V^2}{2g}$$

$$= \left(\frac{900/60}{64.4}\right)^2$$

$$= 3.50 \text{ ft}$$

Note: Because 900 fpm is equivalent to the maximum recommended design velocity (15 fps), it should be readily apparent from the above calculation why velocity head is often ignored, due to its low value when compared to the other head components.

Solution 5.5

Table 5.1 reveals 58.0 ft for one check valve, so two check valves are equivalent to 116.0 ft.

Solution 5.6

Static lift:	10.00 ft
Vertical discharge elevation:	300.00 ft
Shock losses (suction)[(11.0)(2.61)]/100:	0.29 ft
Shock losses (discharge)[(10.6)(3)(8.0)]/100:	2.54 ft
Friction losses (suction)[(25.0)(2.61)]/100:	0.65 ft
Friction losses (discharge)[(500)(10.6)]/100:	53.00 ft
Total dynamic head:	366.48 ft

Solution 5.7

$$E = \frac{(\text{output})}{(\text{input})}$$

$$= \left(\frac{\text{pump output}}{\text{pump input}}\right)\left(\frac{\text{motor output}}{\text{motor input}}\right)$$

$$= \left(\frac{78}{100}\right)\left(\frac{84}{100}\right)$$

$$= \frac{6{,}552}{10{,}000}$$

$$= 65.5\%$$

Solution 5.8

$$HP_B = \frac{(200)(168)(8.33)}{(33{,}000)(0.78)(0.84)}$$

$$= 12.94 \text{ hp}$$

Since 1 hp = 746 W = 0.746 kW (kilowatts), kW = $(HP_B)(0.746)$ = 9.65 kW.

Solution 5.9

To begin, the speed ratio,

$$\frac{(rpm_{New})/(rpm_{Old})(3{,}900)}{(3{,}450)} = 1.13$$

To redraw the curves, the first step is to arbitrarily choose a few points on the pump's head-capacity curve. In this case, the selected points are: (1) Q = 255 gpm, H = 100 ft; (2) Q = 290 gpm, H = 87 ft; and (3) Q = 320 gpm, H = 74 ft.

By reading the corresponding points that are directly beneath (or, sometimes, above) the aforementioned points on the efficiency curve, the values for efficiency and brake horsepower can be determined: (1) E = 77%, HP_B = 8.5; (2) E = 74%, HP_B = 8.9; and (3) E = 66%, HP_B = 9.1.

Original Conditions (3,450 rpm)				New Conditions (3,900 rpm)		
Q	H	HP_B	E	(Q) (SR)	(H) (SR²)	(HP_B) (SR²)
255	100	8.5	0.77	288	128	12.3
290	87	8.9	0.74	328	111	12.8
320	74	9.1	0.66	362	94	13.1

The new head-capacity and brake horsepower curves are plotted on the graph on the next page.

To plot the new efficiency curve, as in the preceding graph, the efficiency equation must be reexamined:

$$E = \frac{QH(8.33)}{HP_B(33{,}000)}$$

If the efficiency equation for the original conditions is as shown here, it follows that the efficiency equation for the new conditions is:

$$E_{New} = \frac{(Q \times SR)(H \times SR^2)(8.33)}{(HP_B \times SR^3)(33{,}000)}$$

134 | MINING ENGINEERING ANALYSIS

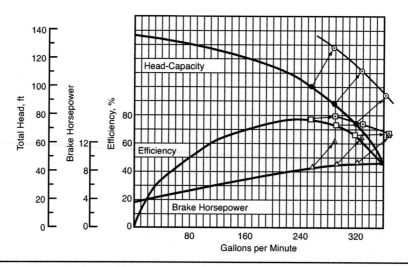

Illustrates Solution to Problem 5.9

Cancellation of the speed ratios in this equation reduces the efficiency equation for the new conditions to the same form as the original equation. This introduces an important concept: **The efficiency does not change for a transposed point!** Thus, the original efficiency values are merely shifted horizontally and aligned with the new head-capacity values, as shown.

Solution 5.10

First, determine the total head losses in the system. Table 5.2 shows the friction-head-loss values for 100 ft of 5-in. pipe at various quantities. Select values at the different quantities, multiply them by 10 (1,000 ft/100 ft), and add 80 (static head) to get the total head losses in the system:

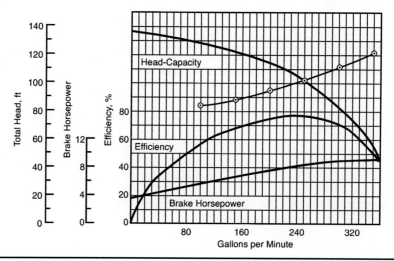

Illustrates Solution to Problem 5.10

Q	$h_f/100$	× (1000/100)	= h_f	+ h_s	= H
100	0.41	10	4.1	80	84.1
150	0.88	10	8.8	80	88.8
200	1.50	10	15.0	80	95.0
250	2.27	10	22.7	80	102.7
300	3.17	10	31.7	80	111.7
350	4.22	10	42.2	80	122.2
400	5.40	10	54.0	80	134.0

Plotting quantity versus total head yields the pipe characteristic curve. The operating point becomes the intersection of the pump and pipe curves, approximately 250 gpm at 103 ft.

Solution 5.11

A common mistake made in a situation such as this is to call the speed ratio $Q_{new}/Q_{original}$. This tendency is wrong because the total head is composed of a variable term (friction head) and a constant term (static head). In situations where a static head is present, the following trial-and-error procedure is recommended:

1. Noting that the present head-capacity curve has a speed ratio of 1.00 and that, because of the aforementioned reason, the speed ratio cannot exceed 1.28 (320 gpm/250 gpm), an arbitrarily chosen speed ratio between these two extremes is selected: 1.20.
2. Because the approximate form of the head-capacity curve is known, only one point, less than Q_{new}, needs to be chosen to initiate the transposition process, for example, 280 gpm at 92 ft.

$$\frac{Q_{original}}{280} \quad \frac{H_{original}}{92}$$

If SR = 1.2, then

$$Q_{new} = \frac{Q_{original} \times 1.2}{336} \quad H_{new} = \frac{H_{original} \times 1.2^2}{132.5}$$

3. Plotting the approximate curve through the point calculated previously indicates that the new operating point would exceed the target of 320 gpm. A smaller speed ratio, 1.1, is then tried.
   If SR = 1.1,

$$Q_{new} = 308, \text{ and } H_{new} = 111.3$$

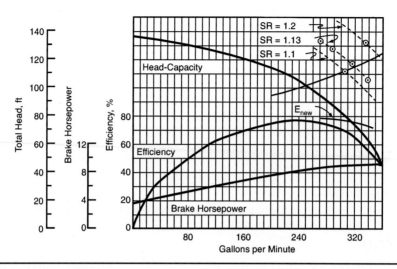

Illustrates Solution to 5.11

Plotting the curve through the point calculated previously indicates that the new operating point falls short of the 320-gpm target. Interpolating between the two approximate curves seems to indicate that the desired speed ratio is approximately 1.13. This time, for accuracy, choose three points on the original curve, including their efficiency values:

$$SR = 1.13$$

Q_o	H_o	E_o	Q_n	H_n	E_n
280	92	0.75	316.4	117.5	0.75
300	83	0.72	339.0	106.0	0.72
240	105	0.78	271.2	134.1	0.78

It should be obvious from the graph that the desired speed ratio is indeed 1.13, the quantity is 320 gpm, the head is 116 ft, and the efficiency, determined from the transposed curve, is 74%.

Solution 5.12

Step 1: Since this is a uniform volume, only four slices need to be passed:
1. Initial water level ($h_s = 25$ ft)
2. $h_s = 100$ ft (required by the problem)
3. $h_s = 200$ ft (required by the problem)
4. Base of the air shaft ($h_s = 300$ ft)

The four slices produce three intervals, as shown in the following figure:

Illustrates Solution to Problem 5.12, Step 1

The quantity of water in each interval can be calculated as follows:

Interval	h ×	π ×	r² ×	Gallons of Water per Cubic Foot	=	Volume (gal)
1	75	3.1416	121	7.48	=	213,255
2	100	3.1416	121	7.48	=	284,339
3	100	3.1416	121	7.48	=	284,339
					Total:	781,933

Step 2: Plot the 5-in. pipe curves onto the pump curves. Because the values required for the gallons per minute exceed those listed in Table 5.2, use Eq. 5.2 and 5.3 to generate the total head (H):

$$H = \left(\frac{L}{100}\right)(0.2083)\left(\frac{100}{C}\right)^{1.85}\left(\frac{q^{1.85}}{d^{4.8655}}\right) + h_s$$

where C is 100, q varies from 0 gpm to 1,500 gpm, d is 5 in., h_s represents the static head at each slice, and L is the pipe length (300 ft).

gpm	H_{start} $h_s = 25$ ft	$h_s = 100$ ft	$h_s = 200$ ft	H_{end} $h_s = 300$ ft
300	34	109	209	309
600	59	134	234	334
900	97	172	272	373
1,200	148	223	323	423
1,500	212	287	387	487

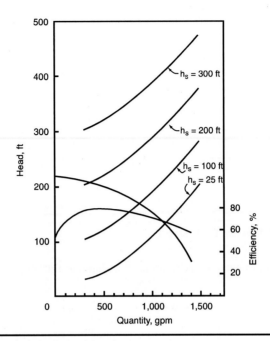

Illustates Solution to Problem 5.12, Step 2

Step 3: Because the pipe curve corresponding to 300 ft of static head does not intersect with the one-stage pump curve, the next step is to see if it intersects with a two-stage curve. It does intersect the two-stage curve, as shown in the following graph. Because the intersection between the pipe curve at the maximum static head value (300 ft) and the two-stage curve also exceeds the inflow rate (400 gpm), it becomes obvious that the two-stage pump can dewater the air shaft.

Step 4: By inspection, the Q_{avg} for each interval is:
1. $(1,360 + 1,240)/2 = 1,300$ gpm
2. $(1,240 + 1,040)/2 = 1,140$ gpm
3. $(1,040 + 775)/2 = 908$ gpm

Step 5: Calculate the time required to pump the water from each interval. (Do not neglect the inflow!) Sum the results to obtain the time that it would take to pump out the air shaft.

Interval 1: $213,255 + 400T = 1,300T$, thus $T = 237$ min
Interval 2: $284,339 + 400T = 1,140T$, thus $T = 384$ min
Interval 3: $284,339 + 400T = 908T$, thus $T = 560$ min
    Total: 1,181 min or 19.7 hr

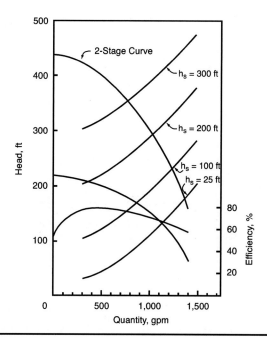

Illustrates Solution to Problem 5.12, Step 3

Step 6: Calculate the representative value for the system's pumping rate.

$$Q \text{ (in gallons per minute)} =$$

$$(\text{standing water}) + \frac{(\text{time in minutes})(\text{inflow rate in gallons per minute})}{(\text{time in minutes})} =$$

$$(781{,}933) + \frac{(1{,}181)(400)}{(1{,}181)} =$$

$$1{,}062 \text{ gpm}$$

Step 7: Reenter the pumping curves to determine the head and efficiency values that correspond to the quantity calculated in step 6.

Entering the curves on the graph on the following page reveals that $H = 286$ ft and $E_p = .70$ (see dashed lines).

Step 8: Calculate kilowatts and total operating costs.

$$kW = \frac{(Q)(8.33)(H)(0.746)}{(33{,}000)(E_p)(E_m)} =$$

$$\frac{(1{,}062)(8.33)(286)(0.746)}{(33{,}000)(0.70)(0.80)} =$$

$$102 \text{ kW}$$

Total cost = (kW)(hr)(cost per kW-hr) = (102)(19.7)($0.08) = $160.75

140 | MINING ENGINEERING ANALYSIS

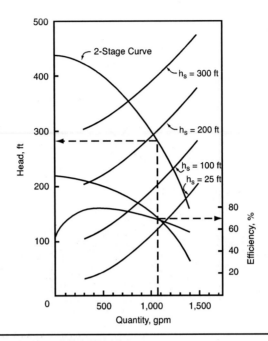

Illustrates Solution to Problem 5.12, Step 7

Solution 5.13

Step 1: Determine the minimum number of slices required.

The minimum number of horizontal slices is five:
1. At the shaft collar
2. At the roof level of the upper drift (h_s = 92 ft)
3. At the floor level of the upper drift (h_s = 100 ft)
4. At the roof level of the lower drift (h_s = 172 ft)
5. At the floor level of the lower drift (h_s = 180 ft)

Five slices produce four intervals, as shown in the following figure.

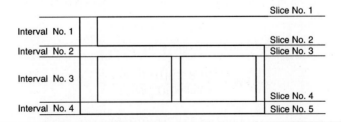

Illustrates Solution to Problem 5.13, Step 1

Step 2: Calculate the volume of the water in each interval.

Interval	Volume (gal)
1	132,127
2	1,806,689
3	133,024
4	1,806,689
Total:	3,878,529

Step 3: Determine the values for the pipe curves (neglect shock losses and velocity head).

Because the values required for the gallons per minute exceed those listed in Table 5.2, use Eq. 5.2 and 5.3 to generate the total head (H):

$$H = \left(\frac{L}{100}\right)(0.2083)\left(\frac{100}{C}\right)^{1.85}\left(\frac{q^{1.85}}{d^{4.8655}}\right) + h_s$$

where C is 100, q varies from 0 gpm to 1,500 gpm, d is 5 in., h_s represents the static head at each slice, and L is the pipe length (180 ft).

		H_{start}			H_{end}	
gpm	h_s = 0 ft	h_s = 92 ft	h_s = 100 ft	h_s = 172 ft	h_s = 180 ft
300	6	98	106	178	186
600	21	113	121	193	201
900	43	135	143	215	223
1,200	74	166	174	246	254
1,500	112	204	212	284	292

Step 4: Plot the pipe curves (shown on the graph on the next page).

Because the curve that represents the minimum rate at the maximum head (550 gpm at 197 ft) exceeds the inflow rate and because all of the curves intersect the single-stage curve, the single-stage pump is adequate.

Step 5: Determine Q_{avg} for each interval.

Interval 1: Q_{avg} = (1,350 + 1,035)/2 = 1,193 gpm
Interval 2: Q_{avg} = (1,035 + 1,000)/2 = 1,018 gpm
Interval 3: Q_{avg} = (1,000 + 620)/2 = 810 gpm
Interval 4: Q_{avg} = (620 + 550)/2 = 585 gpm

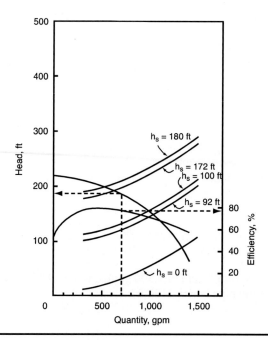

Illustrates Solution to Problem 5.13, Step 4

Step 6:   Determine the time it takes to dewater the mine.

Interval 1: 132,127 + 380T = 1,193T, thus T = 163 min
Interval 2: 1,806,689 + 380T = 1,018T, thus T = 2,832 min
Interval 3: 133,024 + 380T = 810T, thus T = 309 min
Interval 4: 1,806,689 + 380 T = 585 T, thus T = 8,813 min
Total: 12,117 min or 8.4 days

Step 7:   Determine the representative pumping rate that would dewater the mine in the time calculated in step 6.

$$\frac{3{,}878{,}529 + (12{,}117)(380)}{12{,}117} = 700 \text{ gpm}$$

Step 8:   Enter the pumping curves to determine the head and efficiency that relate to 700 gpm.

Enter the curves (as shown by the dashed lines): H = 190 ft, E = 0.78.

Step 9:   Determine the total operating cost.

$$kW = \left[\frac{(700)(8.33)(190)}{(33{,}000)(0.78)(0.82)}\right](0.746) = 39.16 \text{ kW}$$

Operating cost = (39.16)(8.4)(24)($0.08) = $631.57

PUMPING AND DRAINAGE | 143

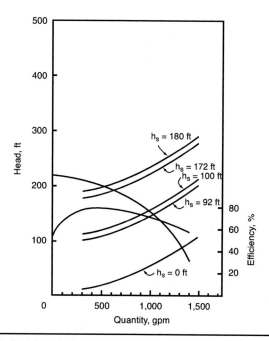

Illustrates Solution to Problem 5.13, Step 8

Solution 5.14

$$\text{Pumping hours required} = \frac{(105)(24)}{(200)} = 12.60 \text{ hr}$$

Solution 5.15

$$\frac{(24 \text{ hr per day})(60 \text{ min per hr})(105 \text{ gpm})}{(8 \text{ pumping hr per day})(60 \text{ min per hr})} = 315 \text{ gpm}$$

Solution 5.16

Original situation:

			Daily Pumping		Costs		
Q	H	E	Hours	kW	Off-peak	On-peak	Daily
200	164	0.78	12.6	9.902	$6.337	$4.555	$10.892

where:

$$kW = \frac{QH(8.33)(0.746)}{(E_m)(E_p)(33,000)}$$

On-peak cost = $(12.6 - 8)(kW)(\$0.10)$

Off-peak cost = $(8)(kW)(\$0.08)$

Speed-change situation:

Q	H	E	Daily Pumping Hours	kW	Costs Off-peak	Costs On-peak	Costs Daily
315	167	0.63	8.0	20.949	$13.407	$0	$13.407

Since $10.892 is less than $13.407, it would be economically advantageous to operate under the conditions existing prior to the speed change.

Solution 5.17

1. Gallons per minute output for the pump:

$$\pi(0.25\,\text{ft})^2 \left(\frac{14\,\text{in.}}{12\,\text{in. per ft}}\right)(30)(7.48\,\text{gal per cu ft}) = 51.4\,\text{gpm}$$

2. Gallons to be removed: $(1.5\,\text{ft})(20\,\text{ft})(220\,\text{ft})(7.48\,\text{gal per cu ft}) = 49{,}368\,\text{gal}$
3. Daily inflow: $(30\,\text{gpm})(60\,\text{min per hr})(24\,\text{hr per day}) = 43{,}200\,\text{gal}$
   Daily outflow: $(51.4\,\text{gpm})(60\,\text{min per hr})(16\,\text{hr per day}) = 49{,}344\,\text{gal}$
   Balance: $49{,}344 - 43{,}200 = 6{,}144\,\text{gal per day}$
4. Days: $49{,}368/6{,}144 = 8.04\,\text{days}$

CHAPTER 6

Mine Power Systems

A mine's power system is the driving force behind all extraction and auxiliary operations, because the production and transportation of mined material and the operation of equipment, such as fans and pumps, all depend on a power source. At one time, compressed air was the primary power source in mining; today, electrical power dominates.

This chapter provides a review of the basic theories of compressed-air and electrical power and gives several examples of applications of power distribution in each system.

COMPRESSED-AIR POWER

Compressed air has been, and should continue to be, an important source of power in mining operations. It has been used to blast coal in conventional mining and is widely used to operate stopers, mucking machines, and other air tools in both coal and hard-rock mines. Such applications demonstrate the suitability of compressed air for applications that require linear motion. Compressed air is also used for its reliability and safety. A compressed-air system is composed of a compressor, a receiver, a distribution network, and the air-operated machines.

The compressor takes in air at normal atmospheric pressure (free air) and compresses it to a higher discharge pressure. The discharge pressure must be high enough to overcome the friction in the distribution system of pipes and hoses and deliver the compressed air to the machines at the pressure recommended by the manufacturer. The most common type of compressor used in mines is the reciprocating compressor in which air is compressed by a piston in a cylinder.

The air receiver is a container or storage tank located in the distribution system between the compressor and the machines. It stores compressed air when the full capacity of the compressor is not being used and gives a more steady flow of air to the machines.

The distribution network consists of pipes, valves, elbows, tees, and hoses that transmit the compressed air from the receiver to the machines in the mine. It is essential that the distribution system be designed with the proper sizes and lengths of pipe, hose, and other components to keep pressure losses well within allowable limits.

The air-operated machines consist primarily of drifters, stopers, pluggers (sinkers), slushers, and several other pieces of equipment.

Compressor Operation

Every compressor is made up of one or more basic elements; a single element, or a group of elements in parallel, comprises a single-stage compressor. Many compression problems involve conditions beyond the practical capability of a single compression stage. Too great a compression ratio (absolute discharge pressure divided by absolute intake pressure) may cause excessive discharge temperatures or other design problems. Consequently, it may become necessary to combine elements or groups of elements in series to form a multistage unit, in which there will be two or more steps of compression. The number of stages commonly used in reciprocating compressors is as follows:

Pressure, psig	Number of Stages
0–150	1
80–500	2
500–2,500	3
2,500–5,000	4

Pressure is frequently measured in pounds per square inch gage (psig), which means the pressure (in pounds per square inch) above barometric pressure as measured by a gage. The sum of barometric pressure and gage pressure is the absolute pressure (in pounds per square inch). The gas is frequently cooled between stages to reduce the temperature and volume entering the subsequent stages, thereby reducing the work required for compression.

The basic reciprocating compression element is a single cylinder that compresses on only one side of the piston (single-acting). A unit that compresses on both sides of the piston (double-acting) consists of two basic single-acting elements operating in parallel in one casing.

The reciprocating compressor uses automatic spring-loaded valves that open only when the proper differential pressure exists across the valves. Intake valves open when the pressure in the cylinder is slightly below the intake pressure. Discharge valves open when the pressure in the cylinder is slightly above the discharge pressure.

Figure 6.1A shows the basic element with the cylinder filled with air at atmospheric pressure. On the corresponding theoretical pressure-volume (pV) diagram, point 1 is the start of compression and both valves are closed.

Figure 6.1B shows the compression stroke during which the piston has moved to the left, thereby reducing the original volume of air with an accompanying rise in pressure. The valves remain closed. The pV diagram shows compression from point 1 to point 2, when the pressure inside the cylinder has reached that in the receiver.

Figure 6.1C shows the piston completing the delivery stroke. The discharge valves are opened just beyond point 2. Compressed air flows out through the discharge valves to the receiver.

After the piston reaches point 3, the discharge valves will close, leaving the clearance space filled with air at discharge pressure. The clearance space is not swept by the piston's movement to protect the valves from damage. During the expansion stroke (Figure 6.1D), both the inlet and discharge valves remain closed and the air trapped in the clearance space increases in volume, causing a reduction in pressure. Pressure continues to fall as the piston moves to the right, until the cylinder pressure drops below the inlet pressure at point 4. The inlet valves then open, allowing air to flow into the cylinder

MINE POWER SYSTEMS | 147

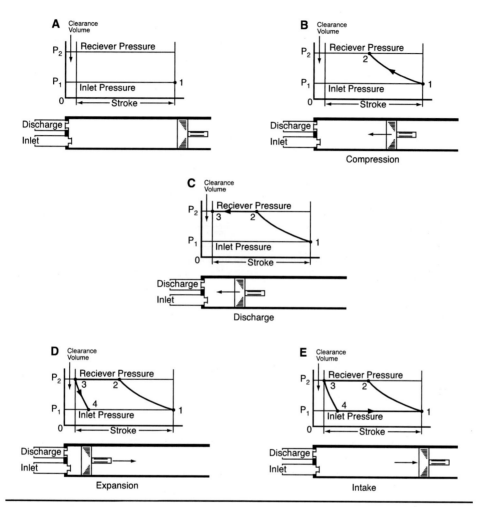

FIGURE 6.1 The various steps in a reciprocating compressor cycle

until the end of the reverse stroke at point 1. This is the intake or suction stroke, which is illustrated in Figure 6.1E. At point 1 on the pV diagram, the inlet valves will close, and the cycle will repeat on the next revolution of the crank.

In a simple two-stage reciprocating compressor, the cylinders are proportioned according to the total compression ratio, $P_2:P_1$ (the ratio of the absolute discharge pressure to the absolute intake pressure), the second stage being smaller because the gas that has already been partially compressed and cooled occupies less volume than at the first stage inlet. Notice in Figure 6.2 that the conditions before the start of compression are points 1 and 5 for the first and second stages, respectively. After compression, the points are 2 and 6; after delivery, they are points 3 and 7. Expansion of air trapped in the clearance spaces as the pistons reverse establishes points 4 and 8. On the intake stroke, the cylinders are again filled at points 1 and 5 as the next cycle commences.

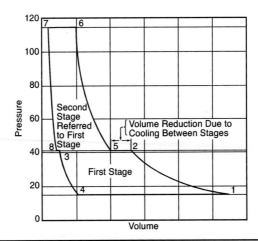

FIGURE 6.2 Combined theoretical indicator card for a two-stage two-element 100-psig positive-displacement compressor

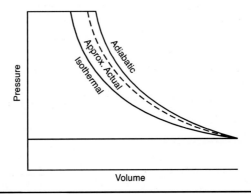

FIGURE 6.3 Adiabatic and isothermal compression chart
Source: Gibbs 1971.

Compressed-Air Theory

Until now, the discussion of compressor operation has dealt primarily with the functions of the components. However, the laws of thermodynamics must be understood to fully appreciate the capabilities of compressors. Thus, it will be necessary to review compressed-air theory before demonstrating applications.

Figure 6.3 shows the relationship between the two theoretical standards and the approximation of the actual compression process on a pV diagram. Note that, without clearance, the volume of free air taken into the piston is equal to the volume displaced by the piston. Compression conforms to the following equation:

$$P_1 V_1^k = P_2 V_2^k \qquad \text{(EQ 6.1)}$$

where k is the ratio of specific heats.

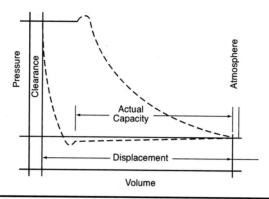

FIGURE 6.4 Effect of cylinder clearance on volumetric efficiency
Source: Gibbs 1971.

However, in theory, compression can follow a number of paths. In the isothermal process (when there is no change in temperature), k in Eq. 6.1 becomes 1.0; in the adiabatic process (when there is no heat added or removed from the system), k becomes 1.395. Neither of these two basic processes is commercially attainable; the actual curve lies somewhere between these two extremes, although it is closer to the adiabatic process.

To calculate the actual work (indicated horsepower) required to compress and deliver a given quantity of gas, it must be remembered that the compression ratio is the ratio of the absolute discharge to the absolute intake pressure ($r = P_2:P_1$). Further, positive-displacement compressors are usually compared to the adiabatic cycle. Thus, for a single-stage compressor, the indicated horsepower (ihp) per 100 cfm is calculated using the following equation:

$$\text{ihp} = 1.542\, P_1 (r^{0.283} - 1) \qquad \text{(EQ 6.2)}$$

The general equation for calculating indicated horsepower per 100 cfm for multistage compressors with intercooling requires a slight alteration of Eq. 6.2:

$$\text{ihp} = X(P_1)1.542(r^{0.283/x} - 1) \qquad \text{(EQ 6.3)}$$

where X is the number of stages.

Compressed-Air System Design

Compressors are used in many real-life situations that have been overlooked in the initial discussions of theory. First, there is cylinder clearance (Figure 6.4), which was also noted in Figure 6.1. Normal clearance, which is the minimum obtainable in a given cylinder, will vary between 4% and 16% for most standard cylinders. Although clearance is of little importance to the average user, because performance is rated on the capacity actually delivered, its effect on capacity should be understood because variation in clearance is used for control and other purposes.

The quantity of air required to operate various tools is rated at sea level; however, the quantity of air actually contained in a cubic foot of free air decreases with increased elevation above sea level. Consequently, corrections for elevations above sea level must be made to capacities. As the quantity of compressed air required to operate the machines remains the same, the cubic feet per minute of free air required increases with increased elevation. Table 6.1 lists the various multipliers for elevations greater than sea

TABLE 6.1 Air consumption multipliers for altitude operation of rock drills based on 80 to 100 psig air pressure

Altitude in feet	0	1,000	2,000	3,000	4,000	5,000	6,000	7,000	8,000	9,000	10,000	12,500	15,000
Multiplier	1.0	1.02	1.05	1.08	1.11	1.14	1.18	1.22	1.26	1.30	1.34	1.46	1.58

Source: Gibbs 1971.

TABLE 6.2 Multipliers for air consumption of rock drills

Number of drills	1	2	3	4	5	6	7	8	9	10	12	15	20	30	40	50	70
Multiplier	1.0	2.0	3.0	4.0	5.0	6.0	6.8	7.5	8.2	9.0	10.5	12.6	16.0	23.5	31.0	38.0	52.5

Source: Gibbs 1971.

TABLE 6.3 Factors for correcting the actual capacity of single-stage compressors at sea level when they are used at higher altitudes (based on 7% cylinder clearance)

Altitude, ft	90 psig Factor	100 psig Factor
Sea level	1.000	1.000
1,000	0.988	0.987
2,000	0.972	0.972
3,000	0.959	0.957
4,000	0.944	0.942
5,000	0.931	0.925
6,000	0.917	0.908
7,000	0.902	0.890
8,000	0.886	0.873
9,000	0.868	0.857
10,000	0.853	0.840
11,000	0.837	—
12,000	0.818	—

Source: Staley 1949.

level; these multipliers yield the necessary capacities when multiplied by the sea-level free-air volumes. Experience shows that all available air-operated tools are very rarely operated simultaneously. Table 6.2 represents the latest consensus with regard to the factors needed to determine the relative air capacity required for operating a number of rock drills from a single compressor system. Because catalog ratings of compressors are based on sea-level conditions, compressor capacities at higher altitudes must be related to sea level. Use the factors in Table 6.3 to make this adjustment; divide the air requirement at the given altitude by the corresponding factor.

Corrections must also be made to pressures at different elevations. Table 6.4 lists the atmospheric pressures at different altitudes. To correct for the effects of altitude on pressure, Eq. 6.4 is used:

TABLE 6.4 Atmospheric pressure at different altitudes

Altitude Above Sea Level, ft	Atmospheric Pressure, psi	Altitude Above Sea Level, ft	Atmospheric Pressure, psi
0	14.69	8,000	10.91
500	14.42	8,500	10.70
1,000	14.16	9,000	10.50
1,500	13.91	9,500	10.30
2,000	13.66	10,000	10.10
2,500	13.41	10,500	9.90
3,000	13.16	11,000	9.71
3,500	12.92	11,500	9.52
4,000	12.68	12,000	9.34
4,500	12.45	12,500	9.15
5,000	12.22	13,000	8.97
5,500	11.99	13,500	8.80
6,000	11.77	14,000	8.62
6,500	11.55	14,500	8.45
7,000	11.33	15,000	8.28
7,500	11.12		

Source: Staley 1949.

TABLE 6.5 Air requirements of representative drilling machines

Type	Hammer Diameter (in.)	Free Air Required (cfm)
Sinker	2³⁄₈	70
Sinker	2¹⁄₂	95
Sinker	2⁵⁄₈	110
Sinker	2³⁄₄	115
Stoper	2⁹⁄₁₆	140
Stoper	2³⁄₄	160
Drifter	2³⁄₄	130
Drifter	3	140
Drifter	3¹⁄₂	180
Drifter	4¹⁄₂	200

Source: Staley 1949.

$$\log P_2 = \log P_1 - \frac{(+ \text{ or } -)h}{122.4(°F + 461)} \quad \text{(EQ 6.4)}$$

where P_2 is the unknown absolute pressure; P_1 is the known absolute pressure; h is the difference in elevation between the two points (+h if increasing in altitude, −h if decreasing in altitude); and °F is the temperature in degrees Fahrenheit (usually 60°F).

Most drilling machines are designed to operate at approximately 90 psig at the tool. Table 6.5 shows the free air required at sea level for various sizes and types of drilling machines. Since most air-operated mining equipment performs best at 90 psig, care must be taken to select a compressor that is capable of delivering air to the receiver at approximately 100 psig, to account for line losses. For example, air-operated tools that work at the design value of 90 psig have a 41% increase in drilling speed over the same tools operated at 70 psig.

To guarantee that the air-operated tools are provided with the correct volume and pressure of air, the distribution system (pipes and flexible hoses) must be properly designed. Normally, line loss should not exceed 5 to 6 psig throughout the entire main line and branch mains. Feed lines (hose) should, therefore, be limited to a 4- to 5-psig line loss if the entire system is to be maintained within the 10-psig guideline. Although tables can be referenced for line loss in pipes, the following equation can be used to determine the pipe diameter that will adequately minimize the line loss:

$$D = \left[\frac{V^2 L}{2,000(P_1^2 - P_2^2)}\right]^{0.20} \quad \text{(EQ 6.5)}$$

where D is the pipe diameter in inches; V is the volume of free air, in cubic feet per minute, that passes through the pipe; L is the pipe length in feet; P_1 is the absolute pressure at the beginning of the pipe; and P_2 is the absolute pressure at the end of the pipe. Table 6.6 shows the various recommended hose lengths and sizes for air-operated tools.

TABLE 6.6 Pressure loss in hose

Hose Length; Inside Diameter	Free Air, cfm	Line Pressure, psig						
		60	80	100	120	150	200	300
50 ft; ¾ in.	60	3.1	2.4	2.0				
	80	5.3	4.2	3.5	2.9	2.4	1.8	1.2
	100	8.1	6.4	5.2	4.5	3.6	2.8	1.9
	120		9.0	7.4	6.3	5.1	3.9	2.7
	140		12.0	9.9	8.4	6.9	5.3	3.6
	160			12.7	10.8	8.9	6.8	4.6
	180				13.6	11.1	8.5	5.8
	200				16.6	13.5	10.4	7.1
	220					16.2	12.4	8.4
50 ft; 1 in.	120	2.7	2.1					
	150	4.1	3.2	2.7	2.3			
	180	5.8	4.6	3.8	3.2	2.6	2.0	1.3
	210	7.7	6.1	5.0	4.3	3.5	2.7	1.8
	240		7.9	6.5	5.5	4.5	3.4	2.3
	270		9.8	8.1	6.9	5.6	4.3	2.9
	300		12.0	9.9	8.4	6.9	5.3	3.6
	330			11.8	10.0	8.2	6.3	4.3
	360			13.9	11.9	9.7	7.4	5.0
	390				13.8	11.3	8.7	5.9
	420				15.9	13.0	10.0	6.8
	450					14.8	11.4	7.7

(table continues on next page)

TABLE 6.6 Pressure loss in hose (continued)

Hose Length; Inside Diameter	Free Air, cfm	Line Pressure, psig						
		60	80	100	120	150	200	300
50 ft; 1¼ in.	200	2.4						
	250	3.7	2.9	2.4	2.0			
	300	5.2	4.1	3.4	2.9	2.3	1.8	1.2
	350	7.0	5.5	4.5	3.8	3.1	2.4	1.6
	400	8.9	7.0	5.8	4.9	4.0	3.1	2.1
	450		8.8	7.3	6.2	5.0	3.9	2.6
	500		10.8	8.9	7.6	6.2	4.7	3.2
	550			10.7	9.1	7.4	5.7	3.9
	600			12.6	10.7	8.7	6.7	4.6
	650			14.6	12.4	10.2	7.8	5.3
	700				14.3	11.7	9.0	6.1
	750					13.3	10.2	6.9
	800					15.0	11.5	7.8
50 ft; 1½ in.	300	2.1						
	400	3.7	2.9	2.4	2.0			
	500	5.6	4.4	3.7	3.1	2.5	1.9	1.3
	600	8.0	6.3	5.2	4.4	3.6	2.8	1.9
	700		8.5	7.0	5.9	4.9	3.7	2.5
	800		10.9	9.0	7.7	6.3	4.8	3.2
	900			11.2	9.5	7.8	6.0	4.1
	1,000			13.6	11.6	9.5	7.3	4.9
	1,100				14.0	11.4	8.8	6.0
	1,200					13.6	10.4	7.1
	1,300					15.8	12.1	8.3

Source: Gibbs 1971.

ELECTRICAL POWER

The tremendous growth in mine mechanization over the past several decades has been facilitated by the increased use of electrical power. Ironically, the design of mine electrical power systems is one area that has been slighted by mining engineers over the years. This is due, perhaps, to a general dislike of electrical theory, the lack of suitable texts dealing with the subject, or a combination of the two factors. However, this situation no longer needs to exist because greater emphasis has been placed on applied research and textbook preparation in the area of mine power systems during the past several years.

For mine-planning purposes, load-flow calculations should be conducted with the four important parameters of voltage (V), current (I), power (P), and power factor (PF) figuring into the output. Fault analysis, which would be treated in a more detailed study, is not covered in this text. The classical approach to load-flow analysis—mathematical solution—is emphasized in the following sections.

Most mine electrical power systems consist of the following seven important components: (1) a-c loads, (2) d-c loads, (3) transmission lines, (4) transformers, (5) rectifiers, (6) power-factor-correction capacitors, and (7) surge capacitors. Figure 6.5 shows a general schematic of a mine electrical power system that indicates (1) the power company's

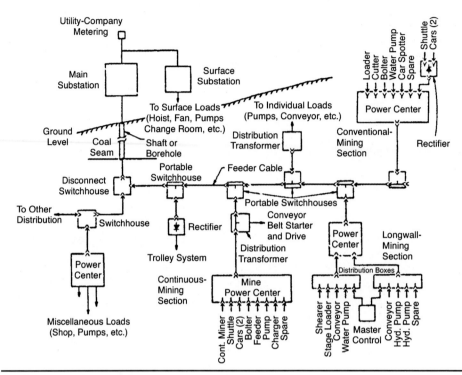

FIGURE 6.5 A radially distributed underground power system
Source: Morley and Novak 1992.

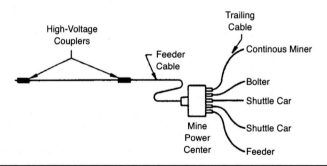

FIGURE 6.6 Utilization in a continuous-miner system
Source: Morley and Novak 1992.

connection at the main mine transformer, (2) the main feeder cable, (3) transmission lines, and (4) panels. The panels contain load centers for transforming the distribution voltage to utilization voltage and rectifying AC to DC. Figure 6.6 illustrates a typical continuous-miner section layout that includes the power center and the loads. The loads are defined in terms of power, voltage, and, in the case of a-c loads, phase angle. Also, the feeder and trailing cables have resistance (R), inductive reactance (L), and capacitive reactance (C); in actual calculations, these three quantities are combined to form an impedance (Z).

This text does not provide a review of the fundamentals of electricity, although knowledge of d-c circuits, steady alternating current, and power is required to complete several problems at the end of the chapter. A standard textbook on electrical engineering fundamentals can provide the reader with the required background, if necessary.

The following sections outline the necessary procedures for designing a preliminary mine electrical power system. Details on selecting power cables and steps to improve power factors are included. The chapter concludes with a discussion of load-flow analysis and a summary of how utilities charge mines for power, emphasizing the need for an efficiently designed mine electrical power system.

Selection of Power Cables

Cables are the connecting link between the various components of the mine electrical power system and the loads. Many properties must be determined for mining cables before a suitable cable can be specified. These include stranding, shielding, breaking strength, weight, insulation, jacket stability, etc. The two most important considerations for mine planning are ampacity and voltage drop.

Ampacity is the current-carrying capacity of a power cable; it is a function of current, ambient temperature, voltage, number of conductors, stranding, duty cycle, and cable geometry. Current (I) requirements can be calculated with the following equation:

$$I = \frac{HP(746)}{(V)(Efficiency)(3^{0.5})(PF)} \quad \text{(EQ 6.6)}$$

where V is the phase-to-phase voltage, HP is the motor horsepower, and PF is the power factor.

The conductor size is then selected from the ampacity tables (Tables 6.7 and 6.8). These ratings are calculated for cables suspended in still air and operating in an ambient temperature of 40°C with a conductor temperature of 90°C. When the ambient temperature differs from 40°C, the values shown in Tables 6.7 and 6.8 should be multiplied by the correction factors listed in Table 6.9. If the cables are to be wound on a reel, the values shown in the ampacity tables should be multiplied by the correction factors shown in Table 6.10. Because the cables are rated under continuous-current conditions, the actual current drawn by mining machinery may be less due to cyclic operation. In these instances, the load factor of the mining equipment should be taken into account.

Voltage drop is the line loss caused by the resistance of the cable to the flow of electricity; it can be likened to the pressure loss in compressed-air systems or the head loss in drainage pipes. The phase-to-phase voltage drop (V) can be calculated using the following equation:

$$V = (3)^{0.5}(I)(Z)[\cos(\theta - \phi)] \quad \text{(EQ 6.7)}$$

where I is the load current in amperes; Z is the single conductor impedance in ohms, which equals $(R^2 + X_L^2)^{0.5}$; θ is the power factor angle; and ϕ is the impedance angle.

Values for R and X_L can be taken from Tables 6.11 and 6.12. Minimization of voltage drop can improve motor life and motor efficiency and can give more efficient utilization of purchased power, making it apparent that voltage drop plays a key role in selecting

TABLE 6.7 Ampacities (conductor temperature: 90°C) for portable power cables in amperes per conductor

Power Conductor Size, AWG or MCM	Single Conductor				Two Conductor, Round and Flat	Three Conductor, Round and Flat	Three Conductor, Round			Four Conductor	Five Conductor	Six Conductor
	0 to 2,000 V Unshielded	2,001 to 8,000 V* Shielded	8,001 to 15,000 V* Shielded	15,001 to 25,000 V* Shielded	0 to 2,000 V	0 to 5,000 V Unshielded	0 to 8,000 V Shielded	8,001 to 15,000 V Shielded	15,001 to 25,000 V Shielded	0 to 2,000 V	0 to 2,000 V	0 to 2,000 V
8	83	—	—	—	72	59	—	—	—	54	50	48
6	109	112	—	—	95	79	93	—	—	72	68	64
4	145	148	—	—	127	104	122	—	—	93	88	83
3	167	171	—	—	145	120	140	—	—	106	100	95
2	192	195	195	—	167	138	159	164	178	122	116	110
1	223	225	225	222	191	161	184	187	191	143	136	129
1/0	258	260	259	255	217	186	211	215	218	165	—	—
2/0	298	299	298	293	250	215	243	246	249	192	—	—
3/0	345	345	343	337	286	249	279	283	286	221	—	—
4/0	400	400	397	389	328	287	321	325	327	255	—	—
250	445	444	440	430	363	320	355	359	360	280	—	—
300	500	496	491	480	400	357	398	—	—	310	—	—
350	552	549	543	529	436	394	435	—	—	335	—	—
400	600	596	590	572	470	430	470	—	—	356	—	—
450	650	640	633	615	497	460	503	—	—	377	—	—
500	695	688	678	659	524	487	536	—	—	395	—	—
550	737	732	—	—	—	—	—	—	—	—	—	—
600	780	779	—	—	—	—	—	—	—	—	—	—
650	820	817	—	—	—	—	—	—	—	—	—	—
700	855	845	—	—	—	—	—	—	—	—	—	—
750	898	889	—	—	—	—	—	—	—	—	—	—
800	925	925	—	—	—	—	—	—	—	—	—	—
900	1,010	998	—	—	—	—	—	—	—	—	—	—
1,000	1,076	1,061	—	—	—	—	—	—	—	—	—	—

Source: Insulated Power Cable Association, 1964; Anaconda Company, 1977.
* These ampacities are based on single isolated cable in air, operated with open-circuit shield.
Note: These ampacities are based on a conductor temperature of 90°C and an ambient air temperature of 40°C.

TABLE 6.8 Ampacities (conductor temperature: 90°C) for three-conductor mine power cables

Conductor Size, AWG or MCM		Ampacities*			
		2,001 to 8,000 V		8,001 to 15,000 V	
Copper	Aluminum	Copper	Aluminum	Copper	Aluminum
6	4	95	95	—	—
4	2	122	124	125	128
2	1/0	159	165	164	168
1	2/0	184	189	187	192
1/0	3/0	211	218	215	221
2/0	4/0	243	251	246	254
3/0	250	279	278	283	281
4/0	350	321	342	325	344
250	400	355	360	359	367
300	450	398	395	401	393
350	500	435	425	438	424
400	—	470	—	473	—
450	—	502	—	504	—
500	—	536	—	536	—

Source: Insulated Power Cable Engineering Association Publication 1964; Anaconda Company, 1977.
*Ampacities are based on an ambient temperature of 40°C and a conductor temperature of 90°C.

TABLE 6.9 Correction factors for insulations rated at 90°C for various ambient temperatures

Ambient Temperature, °C	Corrector Factors
10	1.26
20	1.18
30	1.10
40	1.00
50	0.90

Source: Insulated Power Cable Engineering Association Publication 1964; Anaconda Company, 1977.

TABLE 6.10 Correction factors for cables used with one or more layers wound on a reel

Number of Layers	Multiplying Correction Factors
1	0.85
2	0.65
3	0.45
4	0.35

Source: Insulated Power Cable Engineering Association Publication 1964; Anaconda Company, 1977.

TABLE 6.11 Resistance and reactance of portable power cables

Conductor Size, AWG or MCM	Resistance Ω/1,000 ft		Reactance at 60 Hz Ω/1,000 ft					
	75°C	90°C	2 kv* G – GC G + GC	2 kv SHD – GC	5 kv SHD – GC	8 kv SHD – GC	15 kv SHD – GC	25 kv SHD – GC
8	0.838	0.878	0.034					
7	0.665	0.696	0.033					
6	0.528	0.552	0.032	0.038	0.043			
5	0.418	0.438	0.031	0.036	0.042			
4	0.332	0.347	0.031	0.035	0.040	0.043		
3	0.263	0.275	0.031	0.034	0.039	0.042		
2	0.209	0.218	0.029	0.033	0.038	0.040	0.044	
1	0.165	0.173	0.030	0.033	0.036	0.039	0.042	0.046
1/0	0.128	0.134	0.029	0.032	0.035	0.037	0.040	0.044
2/0	0.102	0.107	0.029	0.031	0.034	0.036	0.039	0.043
3/0	0.081	0.085	0.028	0.030	0.033	0.035	0.038	0.041
4/0	0.065	0.068	0.027	0.029	0.032	0.034	0.036	0.040
250	0.055	0.057	0.028	0.030	0.031	0.033	0.036	0.039
300	0.046	0.048	0.027	0.029	0.031	0.032	0.035	0.038
350	0.039	0.041	0.027	0.029	0.030	0.032	0.034	0.037
400	0.035	0.036	0.027	0.028	0.030	0.031	0.033	0.036
500	0.028	0.029	0.026	0.028	0.029	0.030	0.032	0.035
600	0.023	0.024	0.026	0.027	0.028	0.030	0.032	0.034
700	0.020	0.021	0.026	0.027	0.028	0.029	0.031	0.033
800	0.018	0.019	0.025	0.026	0.028	0.029	0.030	0.033
900	0.016	0.017	0.025	0.026	0.027	0.028	0.030	0.032
1,000	0.014	0.015	0.025	0.026	0.027	0.028	0.030	0.032

Source: Anaconda Company 1977.
* kv (kilovolt).

cables for use in surface and underground mines. In fact, it is recommended that voltage drop be limited to no more than 10% for mine feeder cables.

Power-Factor Correction

As previously mentioned, the power factor is one of the four important parameters of a mine power system. Since it represents the ratio of real power (kilowatt) to apparent power (kilovolt-ampere), it affects the capacity of generating equipment and transformers as well as the required wire sizes and losses in all a-c circuits.

Generally, mine power systems have lagging power factors (I lags V) because of the motors employed, though a leading power factor is just as inefficient. Thus, it is always desirable to raise or correct a low power factor to approach the limiting value of 1.0. For example, if a 500-kW load at 4,160 V and 0.7 PF is corrected to a 0.9 PF, the required current would be reduced by 22%:

TABLE 6.12 Resistance and reactance of mine power feeder cable

Conductor Size, AWG or MCM	Resistance Ω/1,000 ft 90°C	Reactance at 60 Hz Ω/1,000 ft		
		5 kv MP – GC	8 kv MP – GC	15 kv MP – GC
6	0.510	0.041	0.044	
5	0.404	0.040	0.042	
4	0.321	0.038	0.041	
3	0.254	0.037	0.039	
2	0.201	0.036	0.038	0.042
1	0.160	0.035	0.037	0.041
1/0	0.127	0.034	0.035	0.039
2/0	0.101	0.033	0.034	0.038
3/0	0.080	0.032	0.033	0.036
4/0	0.063	0.031	0.032	0.035
250	0.054	0.030	0.031	0.034
300	0.045	0.029	0.031	0.034
350	0.039	0.029	0.030	0.033
400	0.034	0.029	0.030	0.032
500	0.027	0.028	0.029	0.031
600	0.023	0.028	0.029	0.031
700	0.020	0.027	0.028	0.030
800	0.017	0.027	0.028	0.030
900	0.016	0.027	0.027	0.029
1,000	0.014	0.026	0.027	0.029

Source: Anaconda Company 1977.

$$I = \frac{kW(1,000)}{(3)^{0.5}(V)(PF)}$$

at 0.9 PF,

$$I = \frac{(500)(1,000)}{(1.732)(4,160)(0.9)} = 77.1 \text{ amp}$$

at 0.7 PF,

$$I = \frac{(500)(1,000)}{(1.732)(4,160)(0.7)} = 99.1 \text{ amp}$$

A low power factor can be harmful for a variety of reasons:
- It reduces the capacity of the power supply system to transmit real power.
- Each kilowatt of power carries a higher burden of line losses.
- It may depress the voltage, causing a reduction in efficiency.
- It may affect the power bill because many utilities have a penalty for a low power factor.

TABLE 6.13 Power factors of typical a-c loads

Near Unity Power Factor	
Load	Approximate Power Factor
Incandescent lamps	1.0
Fluorescent lamps (with built-in capacitor)	0.95 to 0.97
Resistor heating apparatus	1.0
Synchronous motors (also built for leading power factor operation)	1.0
Rotary converters	1.0
Lagging Power Factor	
Load	Approximate Power Factor
Induction motors (at rated load)	
Split-phase, below 1 hp	0.55 to 0.75
Split-phase, 1 to 10 hp	0.75 to 0.85
Polyphase, squirrel-cage	
High-speed, 1 to 10 hp	0.75 to 0.90
High-speed, 10 hp and larger	0.85 to 0.92
Low-speed	0.70 to 0.85
Wound rotor	0.80 to 0.90
Groups of induction motors	0.50 to 0.85
Welders	
Motor-generator type	0.50 to 0.60
Transformer type	0.50 to 0.70
Arc furnaces	0.80 to 0.90
Induction furnaces	0.60 to 0.70
Leading Power Factor	
Load	Approximate Power Factor
Synchronous motors	0.9, 0.8, 0.7, 0.6, etc. Leading power factors depending on the rated leading power factor for which they are built.
Synchronous condensers	Nearly zero leading power factor. (Output practically all leading reactive kva.)
Capacitors	Zero leading power factor. (Output practically all leading reactive kva.)

Source: Gibbs 1971.

Table 6.13 lists typical power factors for various loads. In mine power systems, it is often difficult to correct power factors by loading all induction motors as fully as possible. Thus, if this is inconvenient, try installing capacitors on each feeder circuit, a procedure followed in most mine power systems.

Calculating the power factor correction value can be simplified by using a tangent table (Table 6.14). The table gets its name from the fact that the ratio of kilovolt-amperes reactive to kilowatts is the tangent of the angle θ (Figure 6.7). Notice that the ratio of kilowatts to kilovolt-amperes, the power factor, is the cosine of the angle θ. Convention dictates that lagging power factors are portrayed by the triangle on the left of

TABLE 6.14   kVAR table

Power Factor	Ratio kvar/kw	Power Factor	Ratio kvar/kw	Power Factor	Ratio kvar/kw
1.00	0.000	0.80	0.750	0.60	1.333
0.99	0.143	0.79	0.776	0.59	1.369
0.98	0.203	0.78	0.802	0.58	1.405
0.97	0.251	0.77	0.829	0.57	1.442
0.96	0.292	0.76	0.855	0.56	1.480
0.95	0.329	0.75	0.882	0.55	1.518
0.94	0.363	0.74	0.909	0.54	1.559
0.93	0.395	0.73	0.936	0.53	1.600
0.92	0.426	0.72	0.964	0.52	1.643
0.91	0.456	0.71	0.992	0.51	1.687
0.90	0.484	0.70	1.020	0.50	1.732
0.89	0.512	0.69	1.049	0.49	1.779
0.88	0.540	0.68	1.078	0.48	1.828
0.87	0.567	0.67	1.108	0.47	1.878
0.86	0.593	0.66	1.138	0.46	1.930
0.85	0.620	0.65	1.169	0.45	1.985
0.84	0.646	0.64	1.201	0.44	2.041
0.83	0.672	0.63	1.233	0.43	2.100
0.82	0.698	0.62	1.266	0.42	2.161
0.81	0.724	0.61	1.299	0.41	2.225

Source: Gibbs 1971.

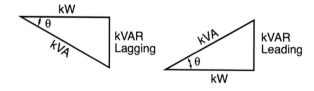

FIGURE 6.7   Power factor triangles
Source: Gibbs 1971.

Figure 6.7, while leading power factors are portrayed by the triangle on the right. Recognizing these relationships, all loads can be broken down into their kilowatt and kilovolt-ampere reactive components. For calculations that involve motors, it is often assumed that 1 hp is approximately equal to 1 kW. The kilowatt and kilovolt-ampere reactive components of two or more loads can be added to determine the overall power factor of the system. To correct lagging power factors, capacitors with their leading power factors can be added to effectively reduce the angle θ, thereby increasing the cosine of θ.

Load-Flow Analysis

Although computer-aided solutions to complex load-flow problems in mine electrical power systems are often warranted, the logic behind the calculations is straightforward and the procedures are not as difficult as they may first appear. This is due to the fact that most mine power systems, no matter how complex, share similarities in design. In general, the procedure is to reduce each panel to an equivalent load. Parallel panel loads are combined to produce an equivalent feeder load, and parallel feeder loads are combined to produce a single equivalent load inby the main mine transformer.

A few of the terms used to discuss load-flow analysis are defined here. The *load factor* is the ratio of the average load to the peak load, both of which occur during the same designated time period. The *demand* is the electrical load for an entire complex averaged over a specific time interval such as 15, 30, or 60 min. Finally, *demand factor* is the ratio of the maximum demand to the total connected load.

Inby the Load Center. Several methods are used to determine the current requirements of a mine power system, but the simplest approach is to use the information provided in a default data table (Table 6.15). This table lists various typical parameters, such as power, effective current, voltage, and power factor, for different models of electrically powered mine equipment. This information greatly simplifies the calculations because mining equipment usually has a cyclic operating nature and does not place a continuous demand on the mine power system.

The first step is to sum the effective currents for section equipment having similar voltages. These total current values will represent the secondary-side requirements of the load center's transformer.

Mine Load Center. The problem at the mine load center is to find the equivalent current on the primary side of the transformer. To do this, the voltage drop inby the secondary side of the transformer is assumed to be less than 10%. In typical practice, voltages of 480 V, 600 V, and 995 V are assumed for the transformer secondary when the loads are 440 V, 550 V, and 950 V, respectively. To cross the transformer, set up the following relationship:

$$1/a = \text{(distribution voltage)}/\text{(secondary voltage)} \qquad \text{(EQ 6.8)}$$

where a is the turns ratio of the secondary. After calculating a, divide it into the required inby current to determine the current on the outby side of the transformer. When completed, the currents from mine equipment of different operating voltages, such as a 950-V continuous miner and a 550-V roof bolter, can be added directly to determine the total current demand on the load center's primary side.

Outby the Mine Load Center. Once the primary-side current requirements have been determined for each operating section, belt drive, etc., they can be added directly to each other in accordance with Kirchhoff's laws, as in any other network. The total current demand of the entire mine can then be expressed all the way back to the secondary of the surface substation.

Power Billing

It cannot be overemphasized that a poorly designed mine electrical power system can be costly in terms of equipment life and efficiency. However, another penalty a poorly

TABLE 6.15 Examples of default data for mining equipment

Equipment	Hp	p (kW)	I_{eff} (amp)	V	PF
Continuous miner	405	181	316.7	550	0.60
	405	181	395.8	440	0.60
	350	147	342.9	550	0.45
	350	147	428.6	440	0.45
	530	214	216.8	950	0.60
	530	214	374.4	550	0.60
	530	214	468.0	440	0.60
Belt drives	75	39	58.5	550	0.70
	75	39	73.1	440	0.70
	150	78	116.9	550	0.70
	150	78	146.2	440	0.70
Feeder breakers	125	65	97.5	550	0.70
	125	65	121.8	440	0.70
Roof bolters					
Single boom	40	15	26.2	550	0.60
	40	15	32.8	440	0.60
Twin boom	80	30	52.5	550	0.60
	80	30	65.6	440	0.60
Section fans	25	11.2	19.6	550	0.60
	25	11.2	24.5	440	0.60
Shuttle cars	135	15	26.2	550	0.60
	135	15	32.8	440	0.60

Source: Morley 1982.

designed system is realized in the form of power costs. Utility companies charge large industrial consumers for more than just the power they used. Table 6.16 shows a typical utility's billing schedule for power service to a mine having a total billing in excess of 5,000 kW per month. Three items in the billing schedule should be noted.

1. The consumer must pay a penalty for reactive kilovolt-ampere demand.
2. The consumer's kilowatt billing is based on the highest capacity demanded.
3. Before power service is provided, an electric service agreement that contains a kilowatt-capacity clause must be signed by all customers. This agreement establishes the minimum monthly bill as well as the size of the utility's service equipment. Thus, it is imperative that the estimation of kilowatt capacity be as accurate as possible; if it is too low, the mine may not have the reliable supply of power it needs, and if it is too high, it can cost the mine a considerable amount of money.

As such, it becomes obvious that a mine power system with a poor power factor or an operating schedule that allows for peaks in the demand rather than scheduling non-time-dependent power consumption during the off-peak period incurs a high penalty.

TABLE 6.16 Typical power billing schedule

Availability

Available to Customers with monthly capacity requirements of 5,000 kilowatts or more that can be served from a 138,000/34,500-V Load Center Substation located within 5 miles of the point of delivery to the Customer. Also available at 12,470 V where the Company elects, at its sole option, to supply Service direct from an adjacent 138,000-V transmission line by a single transformation. Also available to Customers with monthly capacity requirements of 10,000 kilowatts and over, located adjacent to 138,000-V transmission lines. Service shall not be available for Standby or Maintenance Service such as that required for Alternative Generation Facilities. Service will be delivered and metered at 34,500 V or over. An Electric Service Agreement shall be executed.

Monthly Rate

Customer Charge

$124.95

Capacity Charge

$7.42 per kilowatt measured as set forth below under "Determination of Capacity."

Reactive Kilovolt-ampere Charge

$0.40 per reactive kilovolt-ampere of the Customer's reactive kilovolt-ampere capacity requirement in excess of 25% of the Customer's kilowatt capacity.

Energy Charge

All kilowatt-hours at $0.01100 per kilowatt-hour.

Levelized Fuel Rate

Applies to all kilowatt-hours served under this Rate Schedule.

Minimum Charge

Rate Schedule billing.

Determination of Capacity

Kilowatt Capacity

Kilowatt billing capacity (billing demand) during a billing period shall be the highest of:

1. The maximum 30-minute metered kilowatt capacity or
2. For Customers contracting for "off-peak" Service, the billing capacity for any Month shall be the maximum capacity measured during "on-peak" hours plus 25% of the amount that the maximum capacity measured during "off-peak" hours exceeds the maximum "on-peak" capacity subject to provisions set forth below.
   "Off-peak" hours shall be from 10:00 P.M. to 7:00 A.M. daily plus all hours of Sunday.
3. Sixty percent of the maximum instantaneous kilowatt capacity.
   However, the "Billing Capacity" for any Month shall not be less than the highest of:
   a. 5,000 kilowatts, or,
   b. 75% of the kilowatt capacity specified in the Electric Service Agreement, or
   c. 75% of the highest "Billing Capacity" established during the preceding 11 months.

Reactive Kilovolt-ampere Capacity

Reactive kilovolt-ampere capacity is the highest metered demand in reactive kilovolt-amperes established over a 30-minute interval during a billing period.

Kilowatts and reactive kilovolt-amperes will be computed to the nearest whole number.

(table continues on next page)

TABLE 6.16 Typical power billing schedule (continued)

Late Payment Charge

Applies to this Rate Schedule as set forth in Company Rule No. 12 of this tariff.

Electric Service Agreement and Term

An Electric Service Agreement is required for a minimum of 5 years. A new agreement may be required when the Customer increases capacity requirement.

General

Service supplied is subject to the Rules and Regulations Covering the Supply of Electric Service and Rules and Regulations for Meter and Service Installations of the Company as filed with the Commission.

The Company may, at its option, extend Service to the "Point of Service" of a Customer's plant located beyond 5 miles from a Load Center Substation, provided the Customer will pay an additional charge for transmission line facilities in excess of 5 miles. Distances will be computed in poles miles to the nearest tenth of a mile.

Source: Locke 2000.

REFERENCES

Anaconda Co.. 1977. *Mining Cable Engineering Handbook.* Marion, IN: Anaconda Company.
Atlas Copco AB. 1982. *Atlas Copco Manual,* 4th ed. Stockholm, Sweden: Atlas Copco AB.
Gibbs, C.W. 1971. *Compressed Air and Gas Data.* Woodcliff Lake, NJ: Ingersoll-Rand Company.
Krumlauf, H.E. 1963. *The Use of Compressed Air in Small Mines.* Bulletin 172. Tucson, AZ: Arizona Bureau of Mines.
Locke, N.L. 2000. Allegheny Power, Personal Communication.
Morley, L.A. 1982. *Mine Power Systems,* Vol. 1 and 2. Contract No. J0155009. Washington, DC: US Bureau of Mines.
Morley, L.A., and T. Novak. 1992. Electric Power and Utilization. In *SME Mining Engineering Handbook.* Littleton, CO: Society for Mining, Metallurgy, and Exploration, Inc.
Staley, W.W. 1949. *Mine Plant Design.* NY: McGraw-Hill, Inc.

PROBLEMS

Problem 6.1

If each working place of a mine requires four air drills, each with a cylinder diameter of 3 in., that are designed to operate at 90 psig, compute the free-air requirements at an elevation of 4,000 ft if there are five working places.

Problem 6.2

Refer to Problem 6.1. If the working level is 2,000 ft below the surface, compute the compressor rating (single stage) for a surface location.

Problem 6.3

A two-stage compressor with a low-pressure piston displacement of 2,200 cfm is tested by a compressed-air meter that shows a discharge of 220 cfm of compressed air at 90 psig at a temperature of 150°F. The air intake is 60°F, and the compressor is located at an elevation of 3,000 ft. The indicated horsepower of the air cylinders totals 255. What is the volumetric efficiency and adiabatic compression efficiency?

Problem 6.4

Provide a single-stage compressor that will operate a 3-in. drill with a catalog rating of 90 psig when both are at an elevation of 5,000 ft. List all pertinent rating values.

Problem 6.5

Compute the displacement and rating of a single-stage compressor to supply 2,400 cfm of free air at 90 psig at 3,000 ft above sea level. Determine the motor rating needed if the compression efficiency is 88% and the mechanical efficiency is 95%. If the overall motor drive efficiency is 82%, the power cost is $0.045 per kW-hr, and the compressor is operated for 280 days (at 16 hr per day) per year, compute the annual power cost.

Problem 6.6

Determine the gage pressure required of the compressor on the surface (elevation = 2,000 ft) if air is to be delivered to the 1,000-ft level of a mine (disregard friction losses in the pipe) at a gage pressure of 90 psig.

Problem 6.7

If the gage pressure of compressed air delivered to a 2.75-in. stoper drill through three lengths (150 ft) of 1-in. hose is 90 psig, what must be the rated operating pressure of a portable compressor supplying the drill?

Problem 6.8

Calculate the indicated horsepower required to compress 1,600 cfm to 90 psig at an elevation of 1,000 ft. Also determine whether or not a two-stage installation can reduce the indicated horsepower (ihp).

Problem 6.9

A 1,600-ft-long, 10- × 12-ft tunnel is to be driven in hard limestone using a 3-in. drifter operating at 90 psig. Rounds of 8 ft with 24-in. steel changes will be used at an elevation of 4,000 ft. Experience shows that the area of influence for each blasted hole is 5.5 sq ft. Further, time studies indicate that the relationship between hours of drilling and total footage drilled is as follows:

$$\text{Hours} = 0.052 \times (\text{total footage drilled})$$

If each round is to be drilled in 3 hr and a single length of 1-in. hose is to be used, specify the pipe size required and the catalog rating of the compressor.

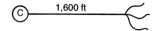

Illustrates Problem 6.9

Problem 6.10

At the furthest extent of operation, two production sections each with seven 3-in. drills are to operate at a minimum gage pressure of 90 psig, as shown in the following diagram. If one length of 1-in. hose is to be used on each drill, what two minimum sizes of pipe will be required to permit a maximum drop of 10 psig in the system? Use a "middle of the road" approach to the compressor rating. Calculate the pressure drops in the entire system. The mine is located at an elevation of 1,000 ft.

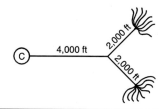

Illustrates Problem 6.10

Problem 6.11

In driving a tunnel, at an elevation of 5,000 ft, that is to eventually reach a distance of 2,000 ft, a contractor must produce 8-ft rounds when the tunnel's cross section is 15 ft × 25 ft in hard limestone. Assuming that he uses 3-in. drifters with two lengths of 1-in. hose and wishes to maintain gage pressures at 90 psig, what size pipe must be used from the compressor station on the surface if a maximum drop of 10 psi can be tolerated between the compressor and drills? He wishes to accomplish drilling of a round in 3 hr. Also, determine the exact compressor rating at the 5,000-ft elevation. Assume that experience has shown that the area of influence of each drill hole in hard limestone is 10.8 sq ft and that one drifter can drill 60 ft of hole in 3 hr.

Problem 6.12

If the gage pressure of a drill jumbo requiring 1,100 cfm of compressed air at the end of 3,000 ft of 4-in. pipe is to be kept at 100 psig, at what pressure must the compressor operate?

Problem 6.13

An exploration tunnel is located 6,000 ft above sea level. Eventually, it will be driven 4,000 ft through hard granite. The tunnel will have a cross section of 9 ft × 10 ft. Assuming that 6-ft rounds will be driven, 3-in. drifters will be used with one length of 1-in. hose, and the gage pressure must be maintained at 90 psig, what minimum pipe size must be used from the compressor on the surface if a maximum drop of 10 psi can be permitted between the compressor and drills? The driller wishes to complete drilling of a round in 4.5 hr. Also, determine the exact compressor rating at the 4,000-ft elevation. Assume that the area of influence for each hole in hard granite is 4.4 sq ft and one drifter can drill 63 ft during 4.5 hr.

Problem 6.14

What is the terminal voltage of a battery when it is connected to a 1-ohm load if the battery's internal voltage is 20 V and the battery's internal resistance is 0.1 ohm?

Problem 6.15

A 300-V d-c shuttle car draws a maximum current of 150 amp. If the resistance of the shuttle car's trailing cable is 0.4 ohms per 1,000 ft, determine the maximum cable length if the rectifier voltage is 300 V and the shuttle car voltage must not fall more than 10% below its rated value.

Problem 6.16

If a mine transformer operates under the conditions shown in the following figure, determine the transformer's secondary voltage and current, the power delivered to the load, the system power factor, and the required kilovolt-amperes of the transformer.

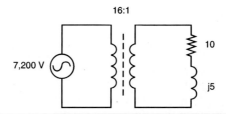

Illustrates Problem 6.16

Problem 6.17

A 250-kVA transformer is at full load with a 0.8 lagging power factor. To correct the power factor to 0.9, a capacitor pack will be added in parallel to the secondary. What are the required kilovolt-amperes reactive of the capacitor pack and how many kilovolt-amperes is the transformer providing at the new power factor?

Problem 6.18

A motor that draws 100 kW at a 0.8 lagging power factor is placed in parallel with another that draws 50 kW at a 0.6 leading power factor. What is their combined power factor?

Problem 6.19

A continuous miner has two 100-hp, 440-V, 60-cps (cycles per second), three-phase squirrel cage induction motors. Calculate the current drawn from the line and the total power factors when:

    a. The first motor is being started.

    b. The second motor is being started and the first motor is running at full speed.

    c. Both motors are being started simultaneously.

    d. Both motors are running at full speed.

The motors have a power factor of 80% lagging at normal full speed and 20% lagging at the time of starting. The motors are started directly on line with full load.

Problem 6.20

A three-phase, 440-V, a-c substation supplies power to the following loads: (1) 60 kVA at 80% power factor lagging and (2) 12 kW at 100% power factor. Calculate the operating power factor of the substation.

Also, a 40-kW synchronous motor will be installed at the substation to improve the station power factor to 95% lagging. Calculate the power factor at which the synchronous motor must operate.

Problem 6.21

A 750-kVA, three-phase, a-c substation operates at 80% lagging power factor. If a synchronous motor is added to the loads, find the power factor at which the motor must operate to improve the station power factor to 95% lagging. Assuming an efficiency of 90%, calculate the horsepower of the largest mechanical load that the synchronous motor can drive in addition to supplying the necessary reactive power.

Problem 6.22

A 530-hp, 950-V a-c continuous miner will be used with a 2/0 type SHD-GC drag cable. Does this cable have adequate ampacity if efficiency is assumed to be 85%? Assume unity power factor.

Problem 6.23

If the cable from the previous problem is 800-ft long and must carry 217 amp, what is the voltage drop?

Problem 6.24

A 135-hp, 550-V a-c shuttle car will be used with a No. 4 three-conductor flat cable. Two layers of the cable will remain on the shuttle car's reel at all times. Does this cable have adequate ampacity if efficiency is assumed to be 80%? Assume unity power factor.

Problem 6.25

Three locomotives, each drawing 300 amp, operate in a system like the one shown in the following figure. In the position depicted, what voltage will be supplied to each locomotive if the trolley wire has a resistance of 0.01 ohms per 1,000 ft and the track has a resistance of 0.015 ohms per 1,000 linear ft? What is the system's efficiency?

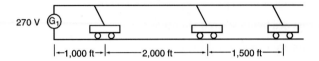

Illustrates Problem 6.25

Problem 6.26

To improve the transmission efficiency in the previous problem, a second conversion unit is placed as shown in the following diagram. If this unit also maintains a constant 270 V, what will the voltages be at each locomotive and what will be the system's transmission efficiency?

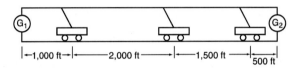

Illustrates Problem 6.26

Problem 6.27

Three locomotives operate between two d-c conversion units that are 3,000 ft apart and maintain a constant voltage of 300 V under all loads. The trolley wire has a resistance of 0.007 ohms per 1,000 ft and the track has a resistance of 0.006 ohms per 1,000 linear ft. Determine the lowest voltage at any locomotive if locomotive A is located midway between the two conversion units and is drawing 1,200 amp, locomotive B is located 1,000 ft from conversion unit B and is drawing 1,000 amp, and locomotive C is located 1,200 ft from conversion unit A and is drawing 800 amp.

Problem 6.28

A continuous-miner section is being planned with the following equipment (including their horsepowers, voltages, and currents):

Equipment	Hp	V	I
Continuous miner	530	950	216.8
Shuttle car 1	135	550	26.2
Shuttle car 2	135	550	26.2
Roof bolter	40	550	26.2
Auxiliary fan	25	550	19.6
Feeder	125	550	97.5
Auxiliary load	10	550	10.0
Charger	50	550	32.0

Determine the current to be supplied to the high side of the load center's transformer. The distribution voltage is 7,200 V.

Problem 6.29

A longwall section is being planned with the following equipment (including their horsepowers and voltages):

Equipment motors	Hp	V
Shearer	400	950
Pump pack 1	50	550
Pump pack 2	50	550
Face conveyor 1	150	550
Face conveyor 2	150	550
Stage loader	50	550

Although the currents are unknown, it is assumed that all motors have load factors of 0.7, except for the motors of the pump pack that are assumed to have load factors of 0.8.

Determine the current to be supplied to the high side of the load center's transformer. The distribution voltage is 7,200 V.

Problem 6.30

In order to determine the voltage drop in the mine power system, the mine planning engineer drew the following electrical schematic, assuming that the "worst-case" situation would occur when both continuous-miner sections and the longwall section were at the farthest point from the portal. With continuous miner section no. 1 being at the most inby point, the voltage drop would be calculated to its load center. If 1/0 cable is projected for use in the panels and 4/0 cable is projected for use in the mains (as shown in the schematic), calculate the voltage drop if the distribution voltage is 7,200 V.

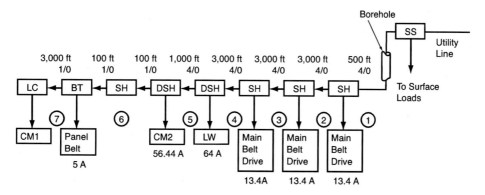

Key: LC-load center, SH-switchhouse, DSH-double switchhouse, SS-surface substation, BT-belt transformer

Illustrates Problem 6.30

Problem 6.31

A large coal mine with a power factor of 80%, consumes 3,500,000 kW-hr during a month with a demand of 9,500 kW. If the current-levelized fuel rate is $0.01013 per kW-hr, calculate the total monthly bill, cost per kilowatt-hour, and monthly load factor using the schedule in Table 6.16.

Problem 6.32

Suppose mine management in the preceding problem signed an electric service agreement with a kilowatt capacity clause of 15,000 kW. If the term of the initial contract is 5 years, determine the excess charges over the entire life of the contract if the power billing schedule in Table 6.16 is used.

PROBLEM SOLUTIONS

Solution 6.1

One drill requires 140 cfm at sea level (see Table 6.5). If there are four drills per working place and five working places, 20 drills will be operating at 4,000 ft. Using Tables 6.1 and 6.2, the free-air requirements can be determined:

$$\text{Free air required} = 140(1.11)(16.0) = 2,486 \text{ cfm}$$

Solution 6.2

For 20 drills at 4,000-ft elevation, $V_c = 2,486$ cfm at 90 psig, where $P_a = 12.68$ psi.

But now the compressor must sit at an elevation of 6,000 ft.

$$\log P_2 = \log P_1 - [(+h)/(122.4(60° + 461°))]$$

$$\log P_2 = \log 102.68 - [(+2,000)/(122.4 (521°))]$$

$$= 2.0115 - 0.0314$$

$$= 1.9801$$

$P_2 = 95.52$, and since P_a at 6,000 ft = 11.77, the compressor's gage pressure is 83.75 psi. The volume of air required is now based on 6,000 ft of elevation:

$$\text{Free air required} = 140(1.18)(16) = 2,643 \text{ cfm}$$

Solution 6.3

Recalling the general gas law (refer to Eq. 1.11 in chapter 1):

$$(p_1V_1)/T_1 = (p_2V_2)/T_2$$

where $p_1 = 13.16$; $p_2 = 90 + 13.16 = 103.16$; $T_1 = 60° + 461° = 521°$; $T_2 = 150° + 461° = 610°$; and $V_2 = 220$ cfm. V_1 becomes 1,470 cfm

$$\text{Volumetric efficiency} = \frac{V_1}{PD} = \frac{1,470}{2,200} = 0.67$$

$$r = \frac{P_2}{P_1} = \frac{103.16}{13.16} = 7.84$$

$$\text{ihp} = 2\,(13.16)(1.542)[7.84^{0.283/2} - 1] = 13.73$$

$$\text{IHP} = 13.73\left(\frac{V_1}{100}\right) = 13.73\left(\frac{1{,}470}{100}\right) = 202 \text{ hp}$$

$$\text{Adiabatic compression efficiency} = \frac{202}{255} = 0.79$$

Solution 6.4

Table 6.5 reveals that a 3-in. drill requires 140 cfm at sea level. To determine the necessary capacity at an elevation of 5,000 ft, Table 6.1 is used to reveal a multiplier of 1.14. Thus, the required free air for the drill at 5,000 ft is 159.6 cfm:

$$140 \text{ cfm } (1.14) = 159.6 \text{ cfm}$$

Because the catalog rating of a compressor is based on operation at sea level, 159.6 cfm at 5,000 ft must be related to sea level, using Table 6.3, as follows:

$$159.6 \text{ cfm} \div 0.931 = 171.43 \text{ cfm}$$

To determine the required gage pressure for the compressor, Eq. 6.4 is used in conjunction with Table 6.4:

$$\log P_2 = \log P_1 - \left[\frac{-h}{(122.4)(60° + 461°)}\right]$$

$$\log P_2 = \log(90 + 12.22) - \left[\frac{-5{,}000}{122.4(521°)}\right]$$

$$\log P_2 = 2.0095 + 0.0784 = 2.0879$$

$$P_2 = 122.46 \text{ psi (absolute)}$$

$$\text{Gage pressure at sea level} = 122.46 - 14.69 = 107.77 \text{ psig}$$

Thus, a compressor with a catalog rating of at least 171.43 cfm and 107.77 psig is required.

Solution 6.5

$$V_c = 2{,}400 \text{ cfm at an elevation of 3,000 ft}$$

Using Table 6.3, the correction factor to refer V_c to sea level is 0.959, so:

$$\text{PD (at sea level)} = \frac{V_c}{0.959} = \left(\frac{2{,}400}{0.959}\right) = 2{,}503 \text{ cfm}$$

$$P_1 = 13.16 \text{ psi}$$

$$P_2 = 13.16 + 90 \text{ psig} = 103.16$$

$$r = \frac{P_2}{P_1} = \frac{103.16}{13.16} = 7.84$$

$$\text{ihp} = 1.542(P_1)(r^{0.283} - 1)$$

$$= 1.542(13.16)(7.84^{0.283} - 1)$$

$$= 16.06$$

$$\text{IHP} = (16.06)\left(\frac{2{,}400}{100}\right) = 385 \text{ hp}$$

$$\text{Motor rating} = \frac{\text{IHP}}{E_{mech} \times E_{compr} \times E_{md}} = \frac{385}{0.88 \times 0.95 \times 0.82} = 562 \text{ hp}$$

Annual cost = 562 hp × 0.746 kW per hp × 280 × 16 × $0.045 = $84,521

Solution 6.6

$$\log P_2 = \log P_1 - \frac{+h}{122.4(60° + 461°)}$$

$$\log P_2 = \log 104.16 - \frac{+1{,}000}{122.4(521°)}$$

$$= 2.018 - 0.016 = 2.002$$

Therefore, $P_2 = 100.46$, and since the atmospheric pressure at 2,000 ft equals 13.66 psi, the gage pressure is 86.80 psig.

Solution 6.7

The volume of air required to operate this drill is 160 cfm (Table 6.5). Refer to Table 6.6. The line loss for a 90-psig drill requiring 160 cfm can be determined as follows:

150 ft of 1-in. hose

	Line Pressure, psig		
Free Air, cfm	80	90	100
150	9.6		8.1
160	11.0	10.1	9.2
180	13.8		11.4

Note: Since hose is usually comes in 50-ft lengths, the values in Table 6.6 must be tripled.

Therefore, the line loss is 10 psi, which is just marginally acceptable if attempting to keep line losses within 10% of the drill's gage pressure.

Solution 6.8

$$P_1 = \text{atmospheric pressure at } 1{,}000 \text{ ft} = 14.16 \text{ psi}$$

$$P_2 = \text{delivery pressure} = 90 + 14.16 = 104.16 \text{ psi}$$

$$r = \frac{P_2}{P_1} = \frac{104.16}{14.16} = 7.36$$

Recalling Eq. 6.2,

$$\text{ihp} = 1.542\, P_1\, (r^{0.283} - 1) = 1.542(14.16)(7.36^{0.283/2} - 1) = 16.58$$

$$\text{IHP} = (16.58)(1{,}600/100) = 265 \text{ hp}$$

Use Eq. 6.3 to determine the indicated horsepower for a two-stage configuration:

$$\text{ihp} = X(P_1)\, 1.542(r^{0.283/X} - 1) = 2(14.16)\, 1.542(7.36^{0.283/2} - 1) = 14.25$$

$$\text{IHP} = (14.25)\left(\frac{1{,}600}{100}\right) = 228 \text{ hp}$$

Percent savings with two stages:

$$\left(\frac{265 - 228}{265}\right) \times 100 = 14\%$$

Solution 6.9

1. Determine the cross-sectional area of the drift:

$$10 \times 12 = 120 \text{ sq ft}$$

2. If the area of influence of each hole is 5.5 sq ft, the number of holes required is:

$$\frac{120}{5.5} = 22 \text{ holes}$$

3. The total length of drilling per round is:

$$22 \times 8 = 176 \text{ ft}$$

4. If drilling will be conducted for 3 hr per shift, the total footage that one drill can produce is:

$$\text{Footage} = \frac{3.0}{0.052} = 58 \text{ ft}$$

Therefore, three drilling units will be required:

$$\frac{176 \text{ ft}}{58 \text{ ft per unit}} = 3 \text{ drilling units}$$

5. Determine the capacity requirements of the drills at 4,000 ft. A 3-in. drifter at 90 psig requires 140 cfm at sea level (Table 6.5). One drill at 4,000 ft has the following volume requirement:

Total cfm = (cfm for 1 drill) × (multiplier from Table 6.1) × (multiplier from Table 6.2) = 140 × 1.11 × 1.0 = 155 cfm

6. Because all three hoses are in parallel, the line loss in one is equal to the loss in all three. Therefore, for one 50-ft length of 1-in. hose, Table 6.6 can be used. Using interpolation, it can be shown that the line loss in the hose is approximately 3.0, well within the established guidelines:

cfm	80 psig	90 psig	100 psig
150	3.2		2.7
155	3.4	3.1	2.8
180	4.6		3.8

7. Equation 6.5 can now be used to determine the pipe diameter:

$$D = \left[\frac{V^2 L}{2,000(P_1^2 - P_2^2)}\right]^{0.20}$$

$$= \left[\frac{(465^2)(1,600)}{2,000(112.68^2 - 105.68^2)}\right]^{0.20}$$

≈ 3 in.

where P_1 is the gage pressure at the drill + acceptable system line loss + atmospheric pressure (90 + 10 + 12.68 = 112.68 psi) and P_2 is the gage pressure at the drill + line loss in the hose + atmospheric pressure.

$$90 + 3 + 12.68 = 105.68 \text{ psi}$$

8. The last step is to take the compressor down to sea level using Eq. 6.4 and Table 6.3:

$$\log P_2 = \log P_1 - \frac{-h}{122.4(60° + 461°)}$$

$$= \log 112.68 - \frac{-4,000}{122.4(521°)}$$

$$= 2.115$$

Therefore, $P_2 = 130.32$, and the compressor's rating is 130.32 − 14.69 = 115.63 psig.

$$\text{Volume of the compressor} = \frac{465.2}{0.942} = 495 \text{ cfm}$$

Solution 6.10

A 3-in. drill requires 140 cfm at sea level. Consequently, at an elevation of 1,000 ft, each drill requires:

$$(140)(1.02)(1.0) = 142.8 \text{ cfm}$$

and each section requires 971 cfm (142.8 × 6.6) and both sections taken together require 1,942 cfm.

To determine the line loss for the mine, start at one drill and move outby along its hose to the junction of the seven section hoses, then along the section pipe to the junction with the other section's pipe, then along the main pipeline.

Hose: Since the line loss is close to 3 psi according to Table 6.6, that value will be used.

Pipes: Assuming that a pressure of 93 psig must be delivered to the hose, 7 psi is left to apportion to the 2,000-ft section line and the 4,000-ft main line.

One approach that can be taken is to assess a 3-psi drop in the section pipe and a 4-psi drop in the main-line pipe, as follows:

Section:

$$D = \left[\frac{971^2(2,000)}{2,000(110.16^2 - 107.16^2)}\right]^{0.20} = 4.29 \text{ in.}$$

Mainline:

$$D = \left[\frac{1,942^2(4,000)}{2,000(114.16^2 - 111.16^2)}\right]^{0.20} = 6.45 \text{ in.}$$

Thus, it appears that a 5-in. line in the panels can be used with a 7-in. line in the mains to handle the requirements for this layout.

Solution 6.11

The cross-sectional area of the tunnel is 375 sq ft. If the area of influence for each hole is 10.8 sq ft, 35 drill holes are required per round. The total amount of drilling per round is 280 ft.

If one drifter can drill 60 ft in 3 hr, five drifters will be required.
One drill at 5,000 ft requires: 140(1.14) = 160 cfm.
Five drills at 5,000 ft require: 140(1.14)(5.0) = 800 cfm.

Line Loss in the Hose

cfm	80 psig	90 psig	100 psig
150	6.4		5.4
160	7.3	6.7	6.1
180	9.2		7.6

Note: The values are twice those shown in Table 6.6 because two lengths of hose are needed per drill.

The line loss in the hose is quite high (6.7 psi). That only leaves 3.3 psi for the line loss in the pipe if the 10-psi maximum drop is to be maintained. Thus, Eq. 6.5 is used:

$$D = \left[\frac{(800)^2(2,000)}{2,000(112.22^2 - 108.92^2)}\right]^{0.20} = 3.88 \text{ in.}$$

Therefore, it appears that a 4-in. pipe will be sufficient.

In turn, the compressor should be rated at 100 psig and 800 cfm at an elevation of 5,000 ft.

$$r = \frac{P_2}{P_1} = \frac{100 + 12.22}{12.22} = 9.18$$

$$\text{ihp} = 1.542\,(12.22)(9.18^{0.283} - 1) = 16.45$$

Thus, IHP = (16.45)(800/100) = 132 hp

Solution 6.12

Rearranging Eq. 6.5,

$$P_1 = \left[\left(\frac{V^2 L}{200 D^5}\right) + P_2^2\right]^{0.5}$$

where V is 1,100 cfm; L is 3,000 ft; D is 4 in.; and P_2 is 100 psig + 14.69 psi = 114.69 psi.

P_1 becomes 122 psi and the gage pressure is:

$$122 - 14.69 = 107.31 \text{ psig}$$

Solution 6.13

The cross-sectional area of the tunnel is 90 sq ft.

> The number of drill holes required: 90/4.4 = 20.45, or 21 holes.
> The footage drilled per round: 21 × 6 = 126 ft.
> Thus, two drifters are required.
> At sea level, one drifter requires 140 cfm.
> At 6,000 ft, one drifter requires 140(1.18) = 165 cfm; two drifters require 330 cfm.

Line Loss in the Hose:

cfm	80 psig	90 psig	100 psig
150	3.2		2.7
165	3.9	3.6	3.3
180	4.6		3.8

A line loss of 3.6 psi leaves 6.4 psi for the line loss in the pipe.

Pipe diameter:

$$D = \left[\frac{330^2(4{,}000)}{2{,}000(111.77^2 - 105.37^2)}\right]^{0.20} = 2.75 \text{ in.}$$

Thus, a 3-in. pipe should be sufficient.

Because the atmospheric pressure at 6,000 ft is 11.77 psi,

$$r = \frac{P_2}{P_1} = \frac{100 + 11.77}{11.77} = 9.5$$

$$\text{ihp} = (1.542)(11.77)(9.5^{0.283} - 1) = 16.17$$

$$\text{IHP} = (16.17)(330/100) = 53.4 \text{ hp}$$

Solution 6.14

A battery can be modeled as an ideal d-c source in series with a resistor to represent internal battery resistance.

$$V_B = V_b \frac{R_{\text{Load}}}{R_S + R_{\text{Load}}} = 20 \frac{(1.0)}{(0.1 + 1.0)} = 18.18 \text{ V}$$

Solution 6.15

$$R_2 = \frac{V_2}{I_2} = \frac{0.9 \times 300}{150} = 1.8 \ \Omega$$

$$V = I(R_1 + R_2)$$

$$300 = 150(1.8 + R_1) = 270 + 150 R_1$$

$$R_1 = 0.2 \ \Omega$$

At 0.4 Ω per 1,000 ft, 500 ft of cable is the maximum length allowable.

Solution 6.16

a. $V_2 = \frac{N_2}{N_1} V_1 = \frac{1}{16}(7{,}200) = 450 \text{ V}$

b. $I_2 = \frac{V_2}{Z_2} = \frac{450 \angle 0°}{11.2 \angle 26.6°} = 40.2 \angle -26.6° \text{ amp}$

c. $P = VI \cos \theta = 450(40.2)(\cos 26.6°) = 16.2 \text{ kW}$

   $Q = VI \sin \theta = 8.1 \text{ kVAR}$

   $S = VI^* = 18.1 \angle +26.6° \text{ kVAR}$

d. $\cos 26.6° = .89$ lagging

e. $S = I_2^2 Z$

$\qquad = (40.2)^2(11.2) = 18.1$ kVA

Solution 6.17

$\qquad P = VI \cos \theta = 250(0.8) = 200$ kW

$\qquad Q = VI \sin \theta = 250(0.6) = 150$ kVAR lagging

At 0.9 lagging, $\theta = 26°$.

$\qquad S' = (200/0.9) = 222$ kVA

The transformer is now providing 222 kVA.

$\qquad Q' = 222 \sin 26° = 97.3$ kVAR lagging

Hence, the capacitor pack's kVAR:

$\qquad = Q - Q'$

$\qquad = 150 - 97.3$

$\qquad = 52.7$ kVAR

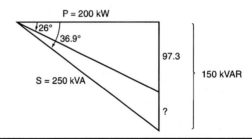

Illustrates Solution to 6.17

Solution 6.18

Motor A

$\qquad P_A = VI \cos \theta = 100$ kW

$\qquad |S_A| = \dfrac{100}{0.8} = 125$ kVA

$\qquad Q_A = VI \sin \theta = 125(0.6) = 75$ kVAR

Motor B

$\qquad P_B = 50$ kW

$\qquad |S_B| = \dfrac{50}{0.6} = 83.3$ kVA

$\qquad Q_B = (83.3)(0.8) = 67$ kVAR

	P	Q
M_A:	100	75
M_B:	50	−67
	150 kW	8 kVAR

$S = P + jQ$

$= 150 + j8 = 150.2 \angle +3.05°$

$pf = \cos \theta = 0.999$ lagging

Solution 6.19

a. First motor is starting:

Motor is starting, hence PF = 20%; we know,

$$\text{Power, watts} = (3)^{0.5} VI \cos \theta$$

substituting the given values,

$$100 \text{ hp} \times 746 = (3)^{0.5} \times 440 \times I \times 0.20$$

or

$$I = \frac{100 \text{ hp} \times 746}{(3)^{0.5} \times 440 \times 0.20} = 489 \text{ amp}$$

Total power factor in this case is 20% lagging.

b. Second motor is starting and first motor is running:

First motor is running, hence its PF = 80%; current taken by this motor:

$$I_1 = \frac{100 \text{ hp} \times 746}{(3)^{0.5} \times 440 \times 0.80} = 122 \text{ amp}$$

Second motor is starting, hence its PF = 20%; current taken by second motor:

$$I_2 = \frac{100 \text{ hp} \times 746}{(3)^{0.5} \times 440 \times 0.20} = 489 \text{ amp}$$

Total current taken by the motors is the vectorial sum of I_1 and I_2. Vectorial sum of I_1 and I_2:

First motor takes I_1, 122 amp. This current lags behind the voltage by $\cos^{-1} 0.80 = 36°54'$. Hence,

Useful current of first motor = $122 \times \cos 36°54'$

$= 122 \times 0.80$

$= 97.6$ amp

Useless current of first motor = $122 \times \sin 36°54'$

$= 122 \times 0.6$

$= 73.2$ amp

Second motor takes I_2, 489 amp. This current lags behind the voltage by $\cos^{-1} 0.20 = 78°30'$. Hence,

Useful current of second motor $= 489 \times \cos 78°30'$

$= 489 \times 0.2$

$= 97.8$ amp

Useless current of second motor $= 489 \times \sin 78°30'$

$= 489 \times 0.98$

$= 479.2$ amp

(Note: The useful current of both motors in this case should be the same, as they are of the same rating, 100 hp each.)

Total current drawn from the line I_L,

$I_L = [(\text{total useful current})^2 + (\text{total useless current})^2]^{0.5}$

$= [(97.6 + 97.8)^2 + (73.2 + 479.2)^2]^{0.5}$

$= [(195.4)^2 + (552.4)^2]^{0.5}$

$= (38{,}181 + 305{,}146)^{0.5}$

$= (343{,}327)^{0.5}$

$= 586$ amps

Total PF $= \dfrac{\text{total useful current}}{\text{total current}}$

$= \dfrac{195.4}{586}$

$= 0.333$, or 33.3% lagging

c. Both motors are being started:

Hence, (from a) each motor takes 424 amp at 20% lagging power factor. Because both currents lag behind the voltage by the same angle,

$$\cos^{-1} 0.20 = 78°30'$$

Total current $= 489 + 489 = 978$ amp

Total power factor is 0.20, or 20% lagging.

d. Both motors are running at normal speed:

Each motor operates at 80% lagging power factor. Hence,

$$\text{Total power} = 2 \times 100 \text{ hp} = 200 \text{ hp}$$

$$\text{Total current} = \frac{200 \text{ hp} \times 746}{(3)^{0.5} \times 440 \times 0.8} = 245 \text{ amp}$$

Solution 6.20

In the accompanying figure, AB represents the 12-kW load unity power factor; triangle BCD represents the 60-kVA load at 80% PF lag; triangle ACD is the resultant of these two loads, hence it represents the 40-kW synchronous motor; and triangle AEF represents the power triangle for the station after power factor correction.

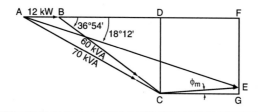

Illustrates Solution to Problem 6.20

With respect to triangle BCD,

$$\text{Load PF, } \cos \theta_L = 0.80; \theta_L = 36°54'$$

Useful power of 60-kVA load,

$\quad$ BD = 60 kVA × PF

$\quad\quad$ = 60 × 0.80

$\quad\quad$ = 48 kW

Reactive power of 60-kVA load,

$\quad$ CD = 60 kVA × sin 36°54'

$\quad\quad$ = 60 × 0.6

$\quad\quad$ = 36 kVAR lagging

In the substation power triangle before power factor correction, triangle ACD:

Total useful power supplied by the substation = AB + BD

$\quad\quad$ = 12 + 48

$\quad\quad$ = 60 kW

Total reactive power supplied by the substation,

$\quad$ CD = 36 kVAR

Total apparent power supplied by the substation,

$$AC = [(\text{total useful power})^2 + (\text{total kVAR})^2]^{0.5}$$
$$= [(60)^2 + (36)^2]^{0.5}$$
$$= 70 \text{ kVA}$$

Substation power factor before correction = (total useful power)/(apparent power)

$$= (60/70) \text{ or } 85.7\% \text{ lagging}$$

After the synchronous motor is installed, triangle AEF:

$$\text{Total useful power} =$$

useful power of loads + useful power supplied by the synchronous motor

or

$$AF = AB + BD + DF$$
$$= 12 + 48 + 40$$
$$= 100 \text{ kW}$$

Desired power factor after improvement.

$$\cos\theta = 0.95; \theta = 18°12'$$

Total reactive power supplied by the station,

$$EF = \text{total useful power} \times \tan 18°\ 12'$$
$$= 100 \times 0.33$$
$$= 33 \text{ kVAR}$$

Reactive power supplied by synchronous motor = substation kVAR before PF correction − substation kVAR after PF correction

or

$$GE = GF - EF$$
$$= 36 - 33 \ [GF = CD]$$
$$= 3 \text{ kVAR leading}$$

In triangle GCE,

$$\tan \varphi_m = \frac{\text{motor reactive power, GE}}{\text{motor useful power, CG}}$$

$$= \frac{3}{40}$$

$$= 0.075$$

$$\phi_m = 4°18'$$

Motor operating power factor,

$$\text{Cos } 4°18' = 0.997 \text{ leading or } 99.7\% \text{ leading}$$

Solution 6.21

In the accompanying figure, triangle AEB and triangle AFC represent the power triangles for the substation before and after the power factor correction, respectively. Triangle BDC represents the synchronous motor.

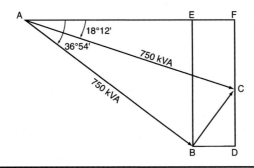

Illustrates Solution to Problem 6.21

With respect to triangle AEB, before power factor correction at 80% lagging:

$$\text{Substation PF, cos } \theta = 0.80; \theta = 36°54'$$

Useful power supplied by the substation,

$\quad$ AE = 750 kVA × cos 36°54'

$\quad\quad$ = 750 × 0.8

$\quad\quad$ = 600 kW

Reactive power supplied by the substation,

$\quad$ EB = 750 kVA × sin 36°54'

$\quad\quad$ = 750 × 0.6

$\quad\quad$ = 450 kVAR lagging

With respect to triangle AFC, after power factor correction to 95% lagging:

$$\text{Station PF, cos } \theta = 0.95; \theta = 18°12'$$

Useful power supplied by the substation,

$\quad$ AF = 750 kVA × cos 18°12'

$\quad\quad$ = 750 × 0.95

$\quad\quad$ = 712.5 kW

Reactive power supplied by the substation,

$$FC = 750 \text{ kVA} \times \sin 18°12'$$
$$= 750 \times 0.312$$
$$= 234 \text{ kVAR lagging}$$

With respect to triangle BDC (representing the synchronous motor), the increase in useful power after the power factor correction,

$$BD = AF - AE$$
$$= 712.5 - 600$$
$$= 112.5 \text{ kW}$$

Reactive power supplied by the synchronous motor,

$$CD = FD - FC$$
$$= 450 - 234 \text{ [FD = EB]}$$
$$= 216 \text{ kVAR leading}$$

Motor phase angle,

$$\theta_m = \tan^{-1}[(216)/(112.5)]$$
$$= \tan^{-1} 1.92$$
$$= 62°30'$$

Operating power factor of the motor,

$$\cos\theta_m = \cos 62°30' \text{ leading}$$
$$= 0.4618 \text{ leading or } 46.18\% \text{ leading}$$

Maximum horsepower of the mechanical load,

Maximum load hp = [(max. useful power of motor)/746] × efficiency

$$= \frac{112.5 \text{ kW} \times 1{,}000}{746} \times 0.90$$

$$= 135.2 \text{ hp}$$

Note that when the station power factor has improved from 80% to 95% lagging, the substation is capable of supplying power to an additional load of 135.2 hp.

Solution 6.22

$$I = \frac{\text{hp}(746)}{(3)^{0.5}(V)(\text{efficiency})(\text{pf})}$$

$$= \frac{(530)(746)}{(3)^{0.5}(950)(0.85)(1.0)}$$

$$= 283 \text{ amp}$$

At 20°C, assuming typical in-mine conditions, the cable ampacity from Table 6.7 (243 amp) multiplied by the correction factor from Table 6.9 (1.18) yields a corrected ampacity of 287 amp. This value would, therefore, be adequate since the calculated value assumes continuous cutting and loading. The actual current value would be less due to the cyclic nature of the operation of the machine. If the continuous miner's load factor is taken into account (typically, 0.5), the average value of cable current will be much less (0.5 × 283 = 142 amp).

Solution 6.23

Using Table 6.11,

$$Z_{2/0} = (R^2 + X_L^2)^{0.5} = [(0.107)^2 + (0.031)^2]^{0.5} = 0.11$$

$$VD = I\,(Z)(3)^{0.5}(L/1{,}000) = 217(0.11)(1.732)(0.8) = 33.07\ V$$

Solution 6.24

$$I = \frac{hp(746)}{(3)^{0.5}(V)(\text{efficiency})(pf)}$$

$$= \frac{(135)(746)}{(3)^{0.5}(550)(0.8)(1.0)}$$

$$= 132.15\ amp$$

Correcting the cable ampacity (Table 6.8) for 20° ambient temperature (Table 6.9) and derating this value for two layers on a reel (Table 6.10) yields:

$$104\ amp\ (1.18)(0.65) = 80\ amp$$

Although it appears that this cable will not work, the cyclic nature of shuttle car operation results in a load factor of approximately 0.4. This means that, on a long-term basis, the cable current averages much less (0.4 × 132.15 = 53 amp). Therefore, this cable should be adequate.

Solution 6.25

The system can be represented by the circuit shown here:

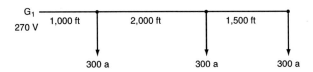

Illustrates Solution to Problem 6.25

The voltages supplied to the three locomotives (L_1, L_2, L_3, as read from left to right) are:

L_1: V = 270 V − (900 amp)(0.025Ω)(1) = 247.5 V

L_2: V = 247.5 V − (600 amp)(0.025Ω)(2) = 217.5 V

L_3: V = 217.5 V − (300 amp)(0.025Ω)(1.5) = 206.25 V

$$E = \frac{output}{input}$$

$$= \frac{(300)(247.5) + (300)(217.5) + (300)(206.25)}{(900)(270)} \times 100\%$$

$$= 82.9\%$$

Solution 6.26

Recalling Kirchhoff's laws for currents and voltages, the system can be represented by the circuit shown here:

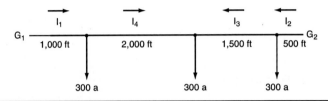

Illustrates Solution to Problem 6.26

$I_1 = 300 + I_4$

$I_2 = 300 + I_3$

$I_3 = 300 - I_4$

$(0.025)(I_1) + 2(0.025)(I_4) = (0.025)(0.5)(I_2) + (1.5)(0.025)(I_3)$

Substituting so that all of the unknowns are in terms of I_3:

$(0.025)(300 + 300 - I_3) + (0.05)(300 - I_3) = (0.0125)(300 + I_3) + (0.0375)(I_3)$

Thus,

$I_3 = 210$ amp

Further:

$I_4 = 300 - I_3 = 90$ amp

$I_2 = 300 + I_3 = 510$ amp

$I_1 = 300 + I_4 = 390$ amp

L_1: V = 270 − (0.025)(390) = 260.25 V

L_2: V = 260.25 − (0.05)(90) = 255.75 V

L_3: V = 255.75 + (0.0375)(210) = 263.63 V

As a check, note that the voltage at the right-hand conversion unit is 270 V:

$$V = 263.63 + (0.0125)(510) = 270 \text{ V}$$

$$E = \frac{\text{output}}{\text{input}}$$

$$= \frac{(300)(260.25 + 255.75 + 263.63)}{(270)(390 + 510)}$$

$$= 96\%$$

Solution 6.27

The described situation can be represented by the circuit shown here:

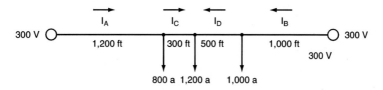

Illustrates Solution to Problem 6.27

Recalling Kirchhoff's laws, the following relationships can be established for currents and voltages:

$$I_a = I_c + 800$$
$$I_b = 1,000 + I_d$$
$$I_c = 1,200 - I_d$$

$$I_a (0.007 + 0.006)(1.2) + I_c(0.013)(0.3) = I_d (0.013)(0.5) + I_b (0.013)(1.0)$$

Dividing through by 0.013 and substituting an equivalent term containing I_c for the unknown terms produces the following relationship:

$$(I_c + 800)(1.2) + I_c(0.3) = (1,200 - I_c)(0.5) + (1,000 + 1,200 - I_c)(1.0)$$

$$I_c = 1,840/3 = 613 \text{ amp}$$

I_d, I_b, and I_a then become:

$$I_d = 1,200 \text{ amp} - 613 \text{ amp} = 587 \text{ amp}$$
$$I_a = 613 \text{ amp} + 800 \text{ amp} = 1,413 \text{ amp}$$
$$I_b = 1,000 \text{ amp} + 587 \text{ amp} = 1,587 \text{ amp}$$

Because locomotive A is the farthest vehicle from the two conversion units, it should have the lowest voltage supplied to any locomotive.

$$(1,413)(0.013)(1.2) + (613)(0.013)(0.3) = (587)(0.013)(0.5) + (1,587)(0.013)(1.0)$$

$$22.05 + 2.39 = 3.82 + 20.62$$

$$24.44 = 24.44$$

Therefore, the lowest voltage is

$$300 - 24.44 = 275.56 \text{ V}$$

Solution 6.28

Because the voltages and the currents are provided, the first step is to sum the currents of similar voltages:

$$950 \text{ V}: \sum I = 216.8 \text{ amp}$$

$$550 \text{ V}: \sum I = 237.7 \text{ amp}$$

Assume that the two secondary windings are set at 995 V and 600 V and determine the turns ratios of the load center's transformer:

$$a_{600} = \frac{7{,}200}{600} = 12.0$$

$$a_{995} = \frac{7{,}200}{995} = 7.24$$

Thus, the high-side current is calculated as follows:

$$I_{600} = (237.7)/(12.0) = 19.8 \text{ amp}$$

$$I_{995} = (216.8)/(7.24) = 29.94 \text{ amp}$$

Total: 49.74 amp

Solution 6.29

When the power factors of the motors are unknown, they are often assumed to be equal to the load factors (up to approximately 0.85). Line currents can then be predicted with the following formula:

$$I = \frac{(\text{load factor})(\text{hp})(746)}{(\text{power factor})(3)^{0.5}(V)}$$

Thus, the currents can be estimated as follows:

$$\text{Shearer } I = \frac{(0.7)(400)(746)}{(0.7)(3)^{0.5}(950)} = 181.4 \text{ amp}$$

$$\text{Pump pack (each motor)} = \frac{(0.8)(50)(746)}{(0.8)(3)^{0.5}(550)} = 39.2 \text{ amp}$$

$$\text{Face conveyor (each motor)} = \frac{(0.7)(150)(746)}{(0.7)(3)^{0.5}(550)} = 117.5 \text{ amp}$$

$$\text{Stage loader} = \frac{(0.7)(50)(746)}{(0.7)(3)^{0.5}(550)} = 39.2 \text{ amp}$$

950 V: $\sum I = 181.4$ amp

550 V: $\sum I = 352.6$ amp

Turns ratios:

$$a_{600} = \frac{7{,}200}{600} = 12.0$$

$$a_{995} = \frac{7{,}200}{995} = 7.24$$

Thus, the high-side current is calculated as follows:

$$I_{600} = (352.6)/(12.0) = 29.38 \text{ amp}$$

$$I_{995} = (181.4)/(7.24) = 25.06 \text{ amp}$$

Total: 54.44 amp

Solution 6.30

The currents to each panel were calculated in the same manner as in the two preceding problems and are given in the schematic. The next step is to determine the impedance (Z) of the feeder cables. Using Table 6.12:

$$Z = (R^2 + X_L^2)^{0.5}$$

$$Z_{1/0} = [(0.127)^2 + (0.035)^2]^{0.5} = 0.132$$

$$Z_{4/0} = [(0.063)^2 + (0.032)^2]^{0.5} = 0.071$$

The magnitude of the voltage drop can be calculated with the following formula:

$$|VD| = |I| \; |Z| \; (3)^{0.5}(L/1{,}000)$$

where L is the cable length in feet.

Breaking the electrical schematic into segments (as indicated by the circled numbers in the figure on the next page), the voltage drops of the segments can be calculated, then added together:

Segment number	Voltage Drop	
7	(51.44)(0.132)(1.732)(3)	= 35.28 V
6	(56.44)(1.32)(1.732)(0.2)	= 2.58 V
5	(112.88)(0.071)(1.732)(1)	= 13.88 V
4	(176.88)(0.071)(1.732)(3)	= 65.25 V
3	(190.28)(0.071)(1.732)(3)	= 70.20 V
2	(203.68)(0.071)(1.732)(3)	= 75.14 V
1	(217.08)(0.071)(1.732)(0.5)	= 13.35 V
	$\sum$ voltage drops =	275.68 V

If a maximum voltage drop of 10% is allowable, this configuration, which represents a voltage drop of only 3.8% (275.68/7,200 V), is acceptable.

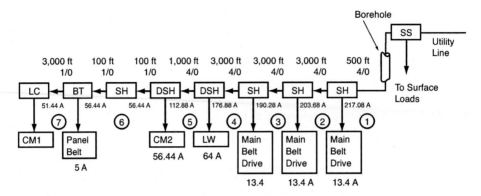

Key: LC-load center, SH-switchhouse, DSH-double switchhouse, SS-surface substation, BT-belt transformer

Illustrates Solution to Problem 6.30

Solution 6.31

Customer charge: $124.95
Capacity charge: 9500 kW @ $7.42 = $70,490.00
Energy charge: 3,500,000 kW-hr @ $0.01100 = $38,500.00
Reactive charge:

$$\cos\theta = \text{power factor} = 0.8 = \frac{kW}{kVA}$$

Therefore,

$$\theta = 36.87°$$

$$\tan\theta = \frac{kVAR}{kW} = 0.75$$

kVAR = kW tan θ = 9,500(0.75) = 7,125 kVAR

Customer's kilowatt demand is 9,500 kW:

(9,500 kW)(0.25) = 2,375

7,125 − 2,375 = 4,750

4,750($0.40) = $1,900.00

Levelized fuel rate: 3,500,000 kW-hr × $0.01013 per kW-hr = $35,455.00
Total monthly bill: $146,469.95
Cost per kilowatt-hour: $146,469.95/(3,500,000 kW-hr) = $0.0418 per kW-hr

Load factor:

$$\frac{\text{kW-hr}}{\text{kW} \times \text{hr per month}} = \frac{3{,}500{,}000}{(9{,}500)(730)} = 0.50$$

A 50% load factor, though seemingly low, is typical for an average coal mine because of widely fluctuating loads. In fact, mine load factors are typically between 20% and 50%.

Solution 6.32

Agreement: 15,000 kW
Billing demand: 15,000 kW × 0.75 = 11,250 kW

Because the mine used only 9,500 kW, it would have to pay for the additional 750 kW as a penalty. Thus, the new kilowatt charge would be calculated as follows:

Kilowatt charge: 11,250 kW @ $7.42 = $83,475.00
Customer charge: $124.95
Energy charge: 3,500,000 kWh @ $0.01100 = $38,500.00
Reactive charge:

$$\cos\theta = \text{power factor} = 0.8 = \frac{\text{kW}}{\text{kVA}}$$

Therefore,

$\theta = 36.87°$

$\tan\theta = \dfrac{\text{kVAR}}{\text{kW}} = 0.75$

kVAR = kW tan θ = 9,500 (0.75) = 7,125 kVAR

Customer's kilowatt demand is 11,250 kW:

(11,250 kW)(0.25) = 2,813

7,125 − 2,813 = 4,312

4,312($0.40) = $1,724.80

Levelized fuel rate: 3,500,000 kW-hr × $0.01013 per kW-hr = $35,455.00
Total monthly bill: $159,179.75
Excess charges: (60 months)($159,179.75 − $146,469.95) = $762,588.00

CHAPTER 7

Hoisting Systems

The transportation of miners, supplies, and mined material from an underground operation is the connecting link between the mine plant and the surface plant. Unless the mine is accessible by a drift or adit, some type of hoisting system is necessary for slopes or shafts.

This chapter deals with the selection of wire ropes and hoists, two of the most important considerations when designing a hoisting system for an underground mine.

WIRE ROPES

A wire rope is designed to transmit forces longitudinally along its axis. It is used in a variety of mining applications because of its flexibility, high tensile strength, and dependability. Shaft hoisting ropes and principal slope haulage ropes are usually custom built; however, proper rope selection for slushers, car spotters, etc., should not be overlooked. Because wire ropes for shafts and slopes represent the lifeline for a mine, they should be made of the highest-quality improved plow steel. Such quality is necessary to deal with the conditions of loading, winding, vibration, abrasion, and corrosion. Tables 7.1 and 7.2 list the specifications for the types of wire ropes typically used for mine haulage applications. The 6 × 19-class haulage rope is used where abrasion is not critical. Where abrasion is critical, such as a slope application, 6 × 7-class haulage ropes are preferred. For other mining applications, manufacturers' handbooks should be consulted.

Selection

A wire rope for a mining situation is selected by simply applying appropriate safety factors to a calculated maximum rope pull and comparing the result to the listed breaking strength of the rope under consideration. Table 7.3 can be used to determine the minimum safety factors for wire ropes.

To determine the maximum rope pull, the following quantities must be known: the rope weight in pounds (x), the conveyance weight in pounds (y), and the weight of the load in the conveyance in pounds (z). These values are used to calculate the load due to gravity, the load due to conveyance friction, and the load due to rope friction. When these loads are added together, as shown in Eq. 7.1, the maximum (static) rope pull (RPS) can be derived:

$$RPS = [(x + y + z)\sin\theta] + [(y + z)(\cos\theta)(0.025)] + [(0.1)(x)(\cos\theta)] \quad \text{(EQ 7.1)}$$

TABLE 7.1 Specifications for 6 × 19-class haulage ropes

Rope Diameter, in.	Approximate Weight, per ft in lbs	Breaking Strength, tons	
		Extra Improved Plow Steel	Improved Plow Steel
5/8	0.66	18.3	16.7
3/4	0.95	26.2	23.8
7/8	1.29	35.4	32.2
1	1.68	46.0	41.8
1 1/8	2.13	57.9	52.6
1 1/4	2.63	71.0	64.6
1 3/8	3.18	85.4	77.7
1 1/2	3.78	101.0	92.0
1 5/8	4.44	118.0	108.0
1 3/4	5.15	136.0	124.0
1 7/8	5.91	155.0	141.0
2	6.72	176.0	160.0
2 1/8	7.59	197.0	179.0
2 1/4	8.51	220.0	200.0
2 1/2	10.5	269.0	244.0

Source: US Steel Supply, 1976, *Wire Rope Engineering Handbook,* US Steel Corporation, Pittsburgh, PA.

TABLE 7.2 Specifications for 6 × 7-class haulage ropes

Rope Diameter, in.	Approximate Weight, per ft in lbs	Breaking Strength, tons	
		Extra Improved Plow Steel	Improved Plow Steel
5/8	0.59	17.5	15.9
3/4	0.84	25.0	22.7
7/8	1.15	33.8	30.7
1	1.50	43.7	39.7
1 1/8	1.90	54.8	49.8
1 1/4	2.34	67.1	61.0
1 3/8	2.84	80.4	73.1
1 1/2	3.38	94.8	86.2

Source: US Steel Supply, 1976, *Wire Rope Engineering Handbook,* US Steel Corporation, Pittsburgh, PA.

TABLE 7.3 Minimum factors of safety (SF) for wire ropes

Vertical Hoist	$SF = 7 - (0.001\ L)$
Slope Hoist	
Koepe Hoist	$SF = 7 - (0.0005\ L)$

Source: US Steel Supply, 1976, *Wire Rope Engineering Handbook,* US Steel Corporation, Pittsburgh, PA.

where θ is the angle of incline and 0.025 and 0.1 are constants determined from experience and used universally.

HOISTS

A mining engineer's concern regarding mine hoist design is to ensure that the motor horsepower of the hoist is sufficient to do its assigned tasks. Obviously, any extended interruption of hoisting service due to improper design can greatly curtail production. To properly size a hoist motor, complete information on the service for which the hoist is intended is needed. The following sections detail the information required for most calculations.

Shaft/Slope Parameters

The following information is necessary for a thorough analysis of hoist-motor design: (1) hoisting layout; (2) shaft/slope inclination; (3) net weight of the loads; (4) weight of skips, cages, and cars; (5) rope size and weight; (6) hoisting distances; (7) drum dimensions; (8) the effective weight of the drums, gears, and sheaves at a given radius from the drum equivalent effective weight (EEW); (9) rope speed; (10) production required; (11) load and dump/caging times; and (12) acceleration and deceleration rates. Although the need for many of these parameters is obvious, those that require further description are clarified in the ensuing paragraphs.

Hoisting Layouts. With each hoisting application, the engineer must determine whether a drum or a Koepe hoist is the better choice; this is not as straightforward as one might expect. One hoist manufacturer has recommended an initial analysis as summarized by Figure 7.1. The hoists can be either single-drum, divided single-drum, double-drum, divided differential-diameter single-drum, single-rope Koepe, or multi-rope Koepe, each with its own advantages and disadvantages. The manufacturer's specifications should be matched to the mining conditions when there is any doubt as to which hoisting layout to choose.

Net Weight of the Loads. When in doubt, the following equation can be used to determine the tonnage of ore that can be handled with each skip:

$$SL = \frac{(D/V) + 0.4V + 12}{(3,600/TPH)} \qquad \text{(EQ 7.2)}$$

where SL is the tonnage of ore that can be handled with each skip, D is the depth in feet, V is the velocity in feet per second, and TPH is the capacity in tons per hour.

Equation 7.2 is based on the assumptions that the sum of the creep and rest times is 12 sec and the acceleration and deceleration rates are each 2.5 fpsps.

Rope Size and Weight. An alternative method for selecting the proper rope size incorporates the following equation and Figure 7.2:

$$d = \left(\frac{SL + SW}{[(K_1/SF) - (K_2D)]/N}\right)^{0.5} \qquad \text{(EQ 7.3)}$$

where d is the rope diameter in inches, SL is the skip load in tons, SW is the skip weight in tons, D is the depth in feet, and N is the number of ropes.

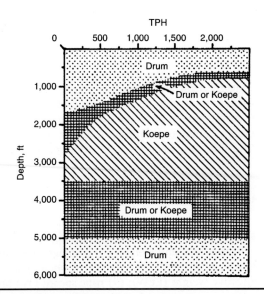

FIGURE 7.1 Areas of applications of drum or Koepe hoists
Source: Gronseth and Hardie 1965.

The values for K_1, K_2, and SF, which is the factor of safety, can be secured from Figure 7.2. A manufacturer's catalog will show the rope weight per foot for the selected diameter.

Drum Dimensions. To find the drum diameter, refer to Figure 7.3 to determine the proper D/d ratio for the hoisting depth. This number, when multiplied by the rope diameter, yields the drum diameter. To find the face width of the drum, substitute the rope diameter into one of the following equations:

$$\text{Single-drum (single layer of rope, balanced hoisting)} = \frac{S(\text{depth})}{\pi(\text{drum diameter})} + 15S \quad \textbf{(EQ 7.4)}$$

$$\text{Double-drum (single layer of rope, balanced hoisting)} = \frac{S(\text{depth})}{\pi(\text{drum diameter})} + 7S \quad \textbf{(EQ 7.5)}$$

where S is the groove pitch (approximately 1.05d).

If the rope is to be layered onto the drums, eliminate the dead-turns values (15S or 7S) and increase the drum diameter by 1.7d for each additional layer of rope.

Effective Weight. The EEW of all rotating parts of the hoist, such as the drums, gears, and sheaves, at a given radius from the drum, must be specified to determine the required horsepower for their acceleration. Such information is often difficult to obtain, but the values shown in Figure 7.4 are recommended by Rexnord, Inc., and are good approximations.

Motor Horsepower for a Drum Hoist

Several variables go into the horsepower calculations for a mine hoist. They are as follows:

D = depth of shaft in feet

RW = rope weight in pounds per foot

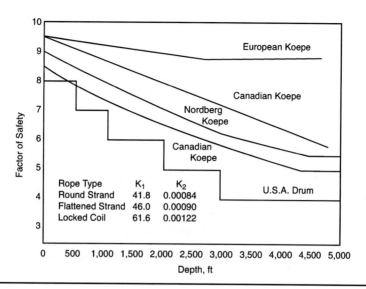

FIGURE 7.2 Minimum factors of safety for wire rope
Source: Gronseth and Hardie 1965.

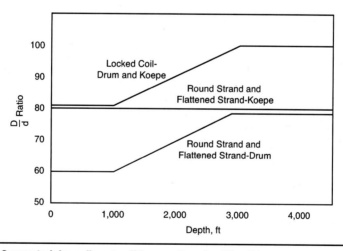

FIGURE 7.3 Suggested drum diameter (D)–rope diameter (d) ratio
Source: Gronseth and Hardie 1965.

V = velocity in feet per second

DT = deceleration time in seconds

SL = skip load in pounds

SW = weight of the skips, cages, and cars in pounds

R = total rope weight in pounds ($D \times RW$)

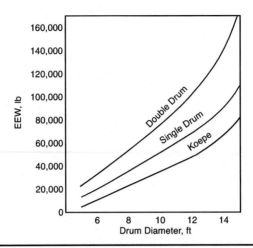

FIGURE 7.4 Approximate equivalent effective weight reduced to rope center for different diameter drums
Source: Gronseth and Hardie 1965.

    E = efficiency expressed as a decimal

  AT = acceleration time in seconds

EEW = equivalent effective weight in pounds

 TSL = total suspended load (EEW + SL + 2SW + 2R)

 SLB = suspended load at the bottom of the shaft [(SL + R) − (V × AT × RW)]

 SLT = suspended load at the top of the shaft [(SL − R) + (V × DT × RW)]

 TFS = full-speed time in seconds

  RT = rest time in seconds

Drum Hoist (Root Mean Square) Horsepower Calculations. There are several separate horsepower calculations that are used to develop the horsepower/time cycle diagram shown in Figure 7.5. They are as follows:

$$HP_1 = \frac{TSL \times V^2}{32.2 \times AT \times 550}$$

$$HP_2 = -\frac{TSL \times V^2}{32.2 \times DT \times 550}$$

$$HP_3 = \frac{(SL + R) \times V}{550}$$

$$HP_4 = \frac{SLB \times V}{550}$$

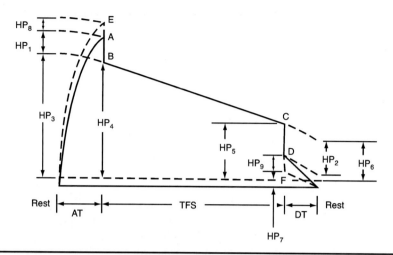

FIGURE 7.5 Horsepower per time cycle for a drum hoist
Source: Gronseth and Hardie 1965.

$$HP_5 = \frac{SLT \times V}{550}$$

$$HP_6 = \frac{(SL - R) \times V}{550}$$

$$HP_7 = \frac{SL \times V}{550} \times \frac{1.0 - E}{E}$$

The horsepower values can be related to give the points that are indicated by letters in Figure 7.5:

    A (peak accelerating horsepower) = $HP_1 + HP_7 + (HP_4 + 2HP_3)/3$

    B (full-speed horsepower at the end of the acceleration period) = $HP_4 + HP_7$

    C (full-speed horsepower at the start of the deceleration period) = $HP_5 + HP_7$

    D (deceleration horsepower) = $HP_2 + HP_7 + (HP_5 + 2HP_6)/3$

Utilizing the peak accelerating horsepower calculated previously, the following two horsepowers can be determined:

    HP_8 (horsepower required to accelerate the motor rotor) = $(0.6A \times 1.2)/AT$

    HP_9 (horsepower required to decelerate the motor rotor) = $-(0.6A \times 1.2)/DT$

Now, points E and F can be determined:

    E (total horsepower required to accelerate the hoist and motor) = $A + HP_8$

    F (total horsepower required to decelerate the hoist and motor) = $D + HP_9$

If the drum hoist has a self-ventilated d-c motor, its root mean square (rms) horsepower will be:

$$\text{rms HP} = \left(\frac{DH}{SV}\right)^{0.5} \tag{EQ 7.6}$$

where:

$$DH = (E^2 \times AT) + \left(\frac{B^2 + C^2 + BC}{3}\right)(TFS) + (F^2 \times DT)$$

and

$$SV = (0.75 \times AT) + (TFS) + (0.75 \times DT) + (0.5 \times RT)$$

If a forced ventilated d-c motor is used, the denominator in the previous equation is replaced by the total cycle time, which includes rest.

If it has an induction motor, its rms horsepower will be:

$$\text{rms HP} = \left(\frac{DH}{IM}\right)^{0.5} \tag{EQ 7.7}$$

where:

$$IM = (0.5 \times AT) + (TFS) + (0.5 \times DT) + (0.25 \times RT)$$

Friction Hoist rms Horsepower Calculations. For a friction (Koepe) hoist, the horsepower calculations are quite similar, except that:

$$R = D \times RW \times 2 \times (\text{number of ropes})$$

and

$$TSL = EEW + SL + 2SW + R$$

With this in mind, the horsepower calculations for Figure 7.6 can be made:

$$HP_1 = \frac{TSL \times V^2}{32.2 \times AT \times 550}$$

$$HP_2 = -\frac{TSL \times V^2}{32.2 \times DT \times 550}$$

$$HP_3 = \frac{SL \times V}{550}$$

$$HP_4 = \left(\frac{SL \times V}{550}\right) \times \left(\frac{1.0 - E}{E}\right)$$

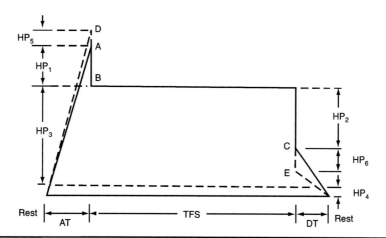

FIGURE 7.6 Horsepower per time cycle for a friction hoist or a drum hoist with a tail rope
Source: Gronseth and Hardie 1965.

The horsepower value can be related to give the points that are indicated by letters in Figure 7.6:

$$A \text{ (peak accelerating horsepower)} = HP_3 + HP_4 + HP_1$$

$$B \text{ (full-speed horsepower)} = HP_3 + HP_4$$

$$C \text{ (total deceleration horsepower)} = HP_3 + HP_4 + HP_2$$

Utilizing the peak accelerating horsepower calculated in the foregoing, the following two horsepowers can be determined:

$$HP_5 \text{ (horsepower required to accelerate the motor rotor)} = \frac{0.75 \times A \times 1.2}{AT}$$

$$HP_6 \text{ (horsepower required to decelerate the motor rotor)} = -\frac{0.75 \times A \times 1.2}{DT}$$

Now, points D and E can be determined:

$$D \text{ (total horsepower required to accelerate the hoist and motor)} = A + HP_5$$

$$E \text{ (total horsepower required to decelerate the hoist and motor)} = C + HP_6$$

If the friction hoist has a self-ventilated d-c motor, its rms horsepower will be:

$$\text{rms HP} = \left(\frac{FH}{SV}\right)^{0.5} \quad \textbf{(EQ 7.8)}$$

where:

$$FH = (D^2 \times AT) + (B^2 \times TFS) + (E^2 \times DT)$$

If it has an induction motor, its horsepower will be:

$$\text{rms HP} = \left(\frac{FH}{IM}\right)^{0.5} \qquad \text{(EQ 7.9)}$$

REFERENCES

Gronseth, J.R., and R. Hardie. 1965. How Modern Mine Hoists Are Selected. *Engineering and Mining Journal.* 166:6:183–190.

Rexnord, Inc. 1983. *Mine Hoists.* Bulletin 453. Milwaukee, WI: Rexnord, Inc.

US Steel Supply. 1983. Wire Rope Catalog. Pittsburg, CA: US Steel Supply.

US Steel Supply. 1976. *Wire Rope Engineering Handbook.* Pittsburg, CA: US Steel Supply.

PROBLEMS

Problem 7.1

A 1.25-in. 6 × 7-class improved plow-steel haulage rope is to be selected for use with a slope hoist. Mineworkers will not be transported by this hoist. From the slope bottom to the top of the headframe, the profile is as follows:

$$2{,}150 \text{ ft @} + 18°$$

If the maximum anticipated load is 40,000 lb, determine the maximum rope pull and whether or not this rope should be selected.

Problem 7.2

An evacuation hoist is to be installed for an underground coal mine. The vertical hoist will evacuate six miners (at one time) every 4 minutes. The depth of overburden is 550 ft. Would a 0.625-in. 6 × 19-class haulage rope be adequate for this application? Assume that the escape capsule weighs 2,000 lb, the total weight of the miners is 1,500 lb, and the maximum rope velocity is 300 fpm.

Problem 7.3

A vertical hoist is to be designed to haul ore at a rate of 500 tph from a depth of 600 ft at an average skip velocity of 1,200 fpm. Determine the required skip load in tons.

Problem 7.4

Use Equation 7.3 to check the rope diameter determined for Problem 7.2.

Problem 7.5

Design a double-drum hoist for a 1,050-ft vertical shaft based on the following conditions:
1. Balanced operation
2. Weight of load = 5.75 tons
3. Weight of skip = 8.25 tons
4. Desired capacity = 400 tph
5. Acceleration time = 10 sec

6. Deceleration time = 10 sec
  7. Rest time per trip = 8 sec
  8. Efficiency = 0.85
  9. An induction motor to be used
  10. A round strand rope to be used

Problem 7.6

What would be the rms horsepower for the hoist in Problem 7.5 if a self-ventilated d-c motor were used?

Problem 7.7

Design a Nordberg Koepe hoist for a 2,000-ft vertical shaft based on the following conditions:
  1. Desired capacity = 300 tph
  2. Weight of load = 13 tons
  3. Weight of skip = 13 tons
  4. Number of hoisting ropes = 4
  5. Flattened strand ropes used for both hoisting and tail ropes
  6. An induction motor used
  7. Efficiency = 90%
  8. Velocity = 12 fps
  9. Acceleration/deceleration rates = 2.5 ft per sec^2
  10. Rest time per trip = 12 sec

Problem 7.8

What would the rms horsepower for the hoist in Problem 7.7 if a self-ventilated d-c motor were used?

Problem 7.9

Design a single-drum hoist for a 1,780-ft long 18° slope based on the following conditions:
  1. Unbalanced operation
  2. Neglect added rope turns around the drum for attachment
  3. Maximum weight of a trip that carries mineworkers = 7.5 tons
  4. Weight of car = 2.5 tons
  5. Acceleration/deceleration times = 20 sec, each
  6. Efficiency = 85%
  7. An induction motor is to be used
  8. A 6 × 7 round strand rope is to be used
  9. Weight of the maximum load = 20 tons
  10. Full-speed time = 110 sec
  11. The hoist drum with two layers of rope
  12. Rest time at the top and bottom = 10 min, each

PROBLEM SOLUTIONS

Solution 7.1

From Table 7.2, the approximate rope weight per foot is 2.34 lb. The maximum pull on the rope must be determined at these two points using information derived from the most inby point of the 18° grade:

$$RPS = \{[(2{,}150)(2.34)] + 40{,}000\}(\sin 18°) + \\ [(40{,}000)(\cos 18°)(0.025)] + \\ \{(0.1)[(2{,}150)(2.34)](\cos 18°)\}$$

$$= 13{,}915 + 951 + 479$$

$$= 15{,}345 \text{ lb (static)}$$

Turning to Table 7.2, the minimum breaking strength for the 1.25-in. plow-steel rope is 122,000 lb. Dividing this value by 15,345 yields the following safety factor:

$$\text{Plow steel: } \frac{122{,}000}{15{,}345} = 7.95$$

Because the fully suspended rope length is 2,150 ft, Table 7.3 indicates that the minimum factor of safety in this situation should be 5.0. Therefore, if 1.25-in. rope is to be selected, improved plow steel exceeds the recommended minimum safety factor.

Solution 7.2

The greatest load on the rope will occur at a depth of 550 ft. Using Tables 7.1 and 7.3, the gravity, conveyance-friction, and rope-friction loads can be determined.

Approximate weight per foot of 0.625-in. 6 × 19-class rope is 0.66 lb.

$$RPS = [(550)(0.66) + (1{,}500 + 2{,}000)]\sin 90° + 0 + 0 = 3{,}863 \text{ lb}$$

Table 7.3 indicates that the minimum factor of safety should be 6.45. Thus, the selected rope should have a breaking strength in excess of 24,916 lb. According to Table 7.1, an improved plow-steel (33,400 lb) 0.625-in. 6 × 19-class rope exceeds the desired breaking strength.

Solution 7.3

Using Equation 7.2,

$$SL = \frac{(D/V) + 0.4V + 12}{(3{,}600/TPH)}$$

$$= \frac{(600/(1{,}200/60)) + 0.4(1{,}200/60) + 12}{(3{,}600/500)}$$

$$= 7 \text{ tons}$$

Solution 7.4

$$d = \left\{\frac{SL + SW}{[(K_1/SF) - (K_2 D)]/N}\right\}^{0.5}$$

$$= \left[\frac{1.75}{((41.8/7) - (0.00084)(550))/1}\right]^{0.5}$$

$$= 0.564 \text{ in.} = 9/16 \text{ in.}$$

This answer compares quite favorably with the size used for Problem 7.2: 5/8 in.

Solution 7.5

$$\text{Trips per hr} = \frac{400 \text{ tph}}{5.75 \text{ tons per trip}}$$

$$= 70 \text{ tph}$$

$$\text{Total time per trip} = \frac{3{,}600 \text{ seconds per hour}}{70 \text{ trips per hour}} = 51.5 \text{ seconds per trip}$$

Full-speed time = 51.5 − 10 − 10 − 8 = 23.5 sec

Assuming constant rates of acceleration and deceleration, the maximum rope speed is:

$$1{,}050/[(10/2) + 23.5 + (10/2)] = 31.34 \text{ fps}$$

$$d = ((SL + SW)/\{[(K_1/SF) - (K_2 D)]/N\})^{0.5}$$

$$= ((5.75 + 8.25)/\{[(41.8/6) - (0.00084)(1{,}050)]/1\})^{0.5}$$

$$= 1.5 \text{ in.}$$

Drum dimensions: From Figure 7.3, D/d = 62; therefore, the drum diameter equals 8 ft ([(1.5)(62)/12]).

Face width of the drum (Eq. 7.4):

$$\frac{S(\text{depth})}{\pi(\text{drum diameter})} + 7S = \frac{(1.05d)(1{,}050)}{\pi(8)} + 7(1.05d)$$

where d is the rope diameter in inches. Thus, the face width is 6.5 ft (76.825/12).

EEW: From Figure 7.4, EEW is approximately 53,000 lb.

Horsepower calculations: The variables for this problem are as follows:

D = 1,050 ft

RW = 3.78 lb (Table 7.1)

V = 31.34 fps

DT = 10 sec

SL = 11,500 lb

SW = 16,500 lb
R = 3,969 lb
E = 0.85
AT = 10 sec
EEW = 53,000 lb
TSL = 105,438 lb
SLB = 14,284 lb
SLT = 8,716 lb
TFS = 23.5 sec
RT = 8 sec

$$HP_1 = \frac{(105{,}438)(31.34)^2}{(32.20)(10)(550)} = 585$$

$$HP_2 = \frac{-(105.438)(31.34)^2}{(32.2)(10)(550)} = -585$$

$$HP_3 = \frac{(11{,}500 + 3{,}969)(31.34)}{550} = 882$$

$$HP_4 = \frac{(14{,}284)(31.34)}{550} = 814$$

$$HP_6 = \frac{(8{,}716)(31.34)}{550} = 497$$

$$HP_6 = \frac{(11{,}500 - 3{,}969)(31.34)}{550} = 429$$

$$HP_7 = \left[\frac{(11{,}500)(31.34)}{550}\right]\frac{(1.0 - 0.85)}{0.85} = 115$$

$$A = 585 + 115 + \left(\frac{814 + 1{,}764}{3}\right) = 1{,}560$$

B = 814 + 115 = 929

C = 497 + 115 = 612

$$D = -585 + 115 + \frac{497 + 858}{3} = -19$$

$$HP_8 = \frac{(0.6)(1{,}560)(1.2)}{10} = 113$$

$$HP_9 = -\frac{(0.6)(1,560)(1.2)}{10} = -113$$

$$E = 1,560 + 113 = 1,673$$

$$F = -19 - 113 = -132$$

$$\text{rms HP} = \left(\frac{DH}{IM}\right)^{0.5} \quad \text{(Note: DH is obtained from Eq. 7.6.)}$$

$$= \left(\frac{42,311,572}{35.5}\right)^{0.5}$$

$$= 1,100 \text{ hp}$$

Solution 7.6

Using Eq. 7.6:

$$\text{rms HP} = \left(\frac{DH}{SV}\right)^{0.5} = \left(\frac{42,311,572}{42.5}\right)^{0.5} = 1,000 \text{ hp}$$

Solution 7.7

Rope diameter:

$$d = \left[\frac{13 + 13}{(46/7.25) - ((0.00090)(2,000)/4)}\right]^{0.5} = 1.2 \text{ in.}$$

Thus, select a 1.25-in. rope.

Check T_1/T_2 ratio, safety factor, and tread pressure: Because of its mode of operation, a Koepe hoist should be checked for proper tension (T_1/T_2) and tread pressure at the hoist drum. T_1/T_2 should be kept at 1.5 or below, while the tread pressure should be in the range of 250 to 270 psi. Thus,

$$\frac{T_1}{T_2} = \frac{SL + SW + (\text{total rope weight}/2)}{SW + (\text{total rope weight}/2)}$$

$$\frac{T_1}{T_2} = 26,000 + 26,000 + \frac{[(2,000)(2.63)(4)(2)]/2}{26,000(21,040)/2} = 1.55$$

where the total rope weight (assuming the weight of flattened strand rope is approximately equal to that of round strand rope) is equal to the following: shaft depth (in feet) × rope weight (in pounds) × number of hoisting ropes × 2 (due to the weight of the tail rope for balance). Although this ratio is slightly high, we will assume that it is sufficient for the purposes of this problem. The ratio could be improved by changing the weight of the skips. Next, check the safety factor. The breaking strength of a 1.25-in. flattened strand rope is assumed to be equal to the 6 × 19 high-strength steel rope of Table 7.1: 64.6 tons. In reality, it may be higher, so a manufacturer's catalog should be consulted.

$$\text{safety factor} = \frac{(\text{breaking strength})(\text{number of ropes})}{(T_1 \text{ in tons})} = \frac{64.6(4)}{36.52} = 7.07$$

When compared to Figure 7.2, this situation appears to be marginally acceptable, especially if the assumed breaking strength of the rope is a conservative value.

Next, use Figure 7.3 to pick the wheel diameter:

$$D/d = 80; \quad 80(1.25) = 100\text{-in. wheel}$$

The tread pressure (tp) is:

$$tp = \frac{T_1 + T_2}{D}(d)(\text{number of ropes}) = \frac{120{,}080}{(100)(1.25)(4)} = 240 \text{ psi}$$

Horsepower calculations: From Figure 7.4, EEW = 25,000 lb.

The acceleration and deceleration times are:

$$AT = DT = \frac{V}{a} = \frac{12 \text{ fps}}{2.5 \text{ ft per sec}^2} = 4.8 \text{ sec}$$

$$R = D \times RW \times 2 \times (\text{number of ropes}) = 2{,}000(2.63)(2)(4) = 42{,}080 \text{ lb}$$

$$TSL = EEW + SL + 2SW + R = 25{,}000 + 26{,}000 + 52{,}000 + 42{,}080 = 145{,}080 \text{ lb}$$

$$HP_1 = \left(\frac{145{,}080 \times 12^2}{32.2 \times 4.8 \times 550}\right) = 246$$

$$HP_2 = -\left(\frac{145{,}080 \times 12^2}{32.2 \times 4.8 \times 550}\right) = -246$$

$$HP_3 = \frac{26{,}000 \times 12}{550} = 567$$

$$HP_4 = \frac{26{,}000 \times 12}{550} \times \frac{1.0 - 0.9}{0.9} = 63$$

$$A = 567 + 63 + 246 = 876$$

$$B = 567 + 63 = 630$$

$$C = 567 + 63 - 246 = 384$$

$$HP_5 = \frac{0.75 \times 876 \times 1.2}{4.8} = 164$$

$$HP_6 = -\frac{0.75 \times 876 \times 1.2}{4.8} = -164$$

$$D = 876 + 164 = 1{,}040$$

$$E = 384 - 164 = 220$$

To calculate the induction motor's rms horsepower, the full-speed time (TFS) must first be determined:

Distance traveled during acceleration = (0.5)(V)(AT)
= (0.5)(12)(4.8)
= 28.8 ft

Distance traveled during deceleration = (0.5)(V)(DT) = 28.8 ft

Total distance traveled (TDT) during AT and DT = 57.6 ft

$$\text{TFS} = \frac{\text{depth} - \text{TDT}}{V} = \frac{2{,}000 - 57.6}{12} = 162 \text{ sec}$$

$$\text{rms HP} = \left(\frac{FH}{IM}\right)^{0.5}$$

$$= \left(\frac{69{,}745{,}502.4}{169.8}\right)^{0.5}$$

$$= 641 \text{ hp}$$

Solution 7.8

Using Eq. 7.8:

$$\text{rms HP} = \left(\frac{FH}{SV}\right)^{0.5}$$

$$= \left(\frac{69{,}745{,}502.4}{175.2}\right)^{0.5}$$

$$= 631 \text{ hp}$$

Solution 7.9

Rope size: Determine the rope size in the manner outlined at the beginning of this chapter; because this slope will probably be used for haulage of workers and materials into and out of the mine but not for mined material, the rope size should be based on the maximum anticipated load (20 + 2.5 = 22.5 tons). Assume that a 1.25-in. 6 × 7-class improved plow-steel rope will be used.

RPS = {[(45,000) + (1,780 × 2.34)](sin 18°)}
        + [(45,000)(cos 18°)(0.025)]
        + [(0.1)(4,165)(cos 18°)]

= 16,659 lb

Turning to Table 7.2, the minimum breaking strength for the rope is 122,000 lb. Dividing this value by 16,659 yields the following safety factor:

$$\frac{12{,}000}{16{,}659} = 7.32$$

This safety factor exceeds the recommended value, which can be determined using Table 7.3. Because the maximum load will only be encountered on rare occasions (dropping/raising heavy equipment, etc.), it appears that this rope will be more than adequate.

Because the acceleration/deceleration times are 20 sec, each, and the full-speed time is 110 sec, the maximum rope speed is:

$$\frac{1,780}{(20/2) + 110 + (20/2)} = 13.7 \text{ fps}$$

Drum dimensions: From Figure 7.3, D/d = 68; therefore, the drum diameter is approximately 7 ft ([(1.25)(68)/(12)]).

The face width of the drum can be determined with Eq. 7.4, and recognizing that the equation must be altered to accommodate layering:

$$\text{Face width in inches} = \frac{(1.05\ d)(\text{rope length})}{\pi D(1.7\ d)}$$

$$= \frac{(1.31)(1,780)}{\pi(7)(2.13)}$$

$$= 50 \text{ in.}$$

Use a 5-ft-wide face with this drum to accommodate any additional required capacity.

EEW: From Figure 7.4, EEW is approximately 30,000 lb.

Critical points: For the horsepower calculations, the static forces encountered at the top and bottom must be determined:

Point	Loads
Bottom	Gravity load = [20,000 + (1,780 × 2.34)](0.3090) = 7,467
	Car friction load = (20,000)(0.95)(0.025) = 475
	Rope friction load = (1,780)(2.34)(0.95)(0.1) = 396
	Total: SLB = 8,338 lb
Top	Gravity load = (20,000 × 0.3090) = 6,180
	Car friction load = (20,000)(0.95)(0.025) = 475
	Rope friction load = 0
	Total: SLT = 6,655 lb

TSL (total suspended load):

$$\text{TSL} = \text{EEW} + \text{SL} + \text{SW} + \text{R} = \text{EEW} + \text{SLB} = 30,000 + 8,338 = 38,338 \text{ lb}$$

RMS HP for a drum hoist—variables:

  TSL = 38,338 lb
   V = 13.7 fps
  AT = 20 sec
  DT = 20 sec
   SL = (15,000 lb)(0.3090) = 4,635 lb
    R = (1,780 ft)(2.34 lb per ft)(0.3090) = 1,287 lb
  SLB = 8,338 lb
  SLT = 6,655 lb

$$HP_1 = \frac{TSL \times V^2}{32.2 \times AT \times 550} = \frac{(38,338)(13.7^2)}{(32.2)(20)(550)} = 20.32$$

$$HP_2 = -\frac{TSL \times V^2}{32.2 \times DT \times 550} = -\frac{(38,338)(13.7^2)}{(32.2)(20)(550)} = -20.32$$

$$HP_3 = \frac{(SL + R) \times V}{550} = \frac{(4,635 + 1,287)(13.7)}{550} = 147.51$$

$$HP_4 = \frac{SLB \times V}{550} = \frac{(8,338)(13.7)}{550} = 207.7$$

$$HP_5 = \frac{SLT \times V}{550} = \frac{(6,655)(13.7)}{550} = 165.77$$

$$HP_6 = \frac{(SL - R)(V)}{550} = \frac{(4,635) - (1,287)(13.7)}{550} = 83.4$$

$$HP_7 = \frac{(SL)(V)}{550} \times \frac{(100 - E)}{E} = \left(\frac{(4,635)(13.7)}{550}\right)\left(\frac{100 - 85}{85}\right) = 20.37$$

$$A = HP_1 + HP_7 + \frac{HP_4 + 2HP_3}{3} = 20.32 + 20.37 + 167.57 = 208.26$$

$$B = HP_4 + HP_7 = 207.7 + 20.37 = 228.07$$

$$C = HP_5 + HP_7 = 165.77 + 20.37 = 186.14$$

$$D = HP_2 + HP_7 + \frac{HP_5 + 2HP_6}{3} = -20.32 + 20.37 + 110.86 = 110.91$$

$HP_8 = [(0.6A)(1.2)]/AT = 149.9/20 = 7.5$

$HP_9 = [(-0.6A)(1.2)]/DT = -149.9/20 = -7.5$

$$E = A + HP_8 = 208.26 + 7.5 = 215.8$$

$$F = D + HP_9 = 110.91 + (-7.5) = 103.41$$

rms HP = $\{DH/[(0.5 \times AT) + (TFS) + (0.5 \times DT)]\}^{0.5}$

Note: Because the rest time for this operation is not truly a part of the hoist's cycle time, it is neglected in the foregoing equation.

rms HP = $(45,201)^{0.5}$

= 213 hp

The next largest standard hoist motor with an rms HP in excess of 213 would be selected.

CHAPTER 8

Rail and Belt Haulage Systems

Rail systems, conveyor belt systems, or a combination of the two are commonly used to transport personnel, supplies, and mined material in an underground or surface mine. Typically, large underground mines use rail haulage from the mining sections to the hoist for supply and personnel transport are used with conveyor belts to transport mined material. Small underground mines use conveyor belts throughout. The size and layout of surface mines also contribute to variations in the way these haulage systems are applied.

The following sections address the typical components mining engineers must consider when designing these haulage systems. For rail, these include rail selection and turnout design, locomotive tractive-effort calculation, and motor sizing. Belt haulage calculations include belt width and speed specifications, tension and takeup requirements, idler spacings, and motor sizing.

TRACK LAYOUTS

The adequacy of mine track layouts depends on such factors as rail weight (Table 8.1); suitable ties; a well-drained roadbed; and properly designed curves, grades, and track alignment. As long as the installation conforms to the manufacturer's recommendations, good service can be expected from the rail system. Invariably, mining engineering calculations for track layouts deal almost exclusively with the design of turnouts. As such, the following section provides the necessary information for computing the theoretically correct dimensions of turnouts.

Turnout Design

Figure 8.1 shows a typical turnout with components identified; Figure 8.2 identifies the variables used to determine the turnout dimensions.

The variables shown in Figure 8.2 are as follows:

- A: distance from the actual point of a frog to its toe
- B: thickness of the frog point
- C: length of the curved closure rail
- D: length of the straight closure rail

MINING ENGINEERING ANALYSIS

TABLE 8.1 Suggested rail weights

	Weight of Rail, lb per yard	
Weight of Locomotive, tons	Four-wheel Locomotive	Six-wheel Locomotive
6	30	—
8	30	—
10	40	30
13	60	40
15	60	40
20	60	60
25	70	60
30	80	70
35	85	80
40	90	85
50	100	90

Source: Bethlehem Steel Company 1958.

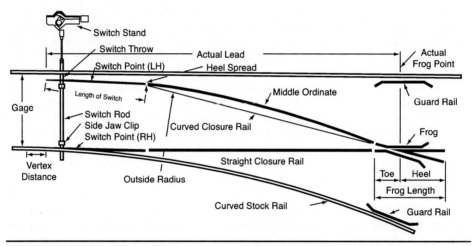

FIGURE 8.1 Typical rail turnout with components indicated
Source: Bethlehem Steel Company 1958.

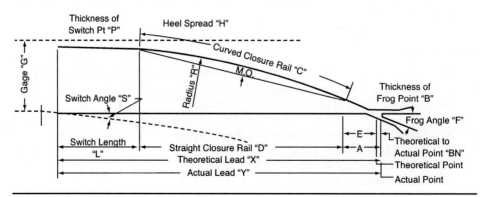

FIGURE 8.2 Variables in turnout calculations
Source: Bethlehem Steel Company 1958.

E: distance from the theoretical point of the frog to its toe
F: frog angle
G: gage
H: heel spread of the switch
L: length of switch
MO: middle ordinate
N: frog number
P: thickness of switch point
R: radius of curved closure rails
S: switch angle
X: theoretical lead
Y: actual lead

The formulas used by the mining engineer include:

$$\text{Switch angle (S)} = \arcsin\left(\frac{H-P}{L}\right) \quad \text{(EQ 8.1)}$$

$$\text{Theoretical toe distance (E)} = A - (B \times N) \quad \text{(EQ 8.2)}$$

(Note: If a cast frog is used, add 10 in. to the value of E.)

$$\text{Theoretical lead (X)} = (L \cos S) + \left(\frac{G\text{-}H\text{-}E \sin F}{\tan 0.5(F+S)}\right) + E \cos F \quad \text{(EQ 8.3)}$$

$$\text{Actual lead (Y)} = X + BN \quad \text{(EQ 8.4)}$$

$$\text{Radius (R)} = \frac{G\text{-}H\text{-}E \sin F}{2 \sin 0.5(F+S) \sin 0.5(F-S)} \quad \text{(EQ 8.5)}$$

$$\text{Length of straight closure rail (D)} = X\text{-}L\text{-}E \quad \text{(EQ 8.6)}$$

(Note: If a cast frog is used, add 10 in. to the value of D.)

$$\text{Length of curved closure rail (C)} = R(F-S) \quad \text{(EQ 8.7)}$$

(Note: F and S must be converted to radians for this formula. Also, if a cast frog is used, add 10 in. to the value of C.)

$$\text{Middle ordinate (MO)} = R - R \cos 0.5(F-S) \quad \text{(EQ 8.8)}$$

LOCOMOTIVE TRACTIVE-EFFORT CALCULATIONS

Mine locomotives for rail haulage are selected based on the weight of the prime mover's horsepower. Figure 8.3 shows the relationship between a mine locomotive and its trailing loads. The tractive effort, which is exerted by the locomotive at the rim of its driving wheels, is a function of its weight and its ability to adhere to the track. Table 8.2 shows adhesive values for locomotives under various conditions; these decimals represent the portions of the prime mover's weight available to haul the trip. Drawbar pull is the portion of the locomotive's tractive effort, in pounds, that remains after the locomotive's resistance has been subtracted.

The locomotive's horsepower can be determined from the following relationship:

$$HP = \frac{(TE)(V)}{375(E)} \quad \text{(EQ 8.9)}$$

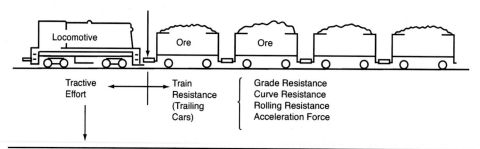

FIGURE 8.3   Relationship between a locomotive and its trailing load
Source: Brauns and Orr 1968.

TABLE 8.2   Adhesive values for locomotives

	Unsanded Rails	Sanded Rails
Clear dry rails, starting and accelerating	0.30	0.40
Clear dry rails, continuous running	0.25	0.35
Clear dry rails, locomotive braking	0.20	0.30
Wet rails	0.15	0.25

Source: Bethlehem Steel Company 1958.

where TE is the available tractive effort in pounds, V is the velocity in miles per hour, and E is the transmission efficiency.

Forces Needed to Move the Rolling Stock

To move a trip, a mine locomotive must be capable of overcoming the following resistances for both itself and the mine cars: (1) rolling resistance, (2) curve resistance, (3) grade resistance, and (4) acceleration or deceleration.

The rolling resistance of the entire train is equal to the weight, in tons, of the locomotive and mine cars (including the payload, if any) multiplied by a frictional coefficient, in pounds per ton. When roller bearings are used, the coefficient is 20 lb per ton of weight; the coefficient for plain bearings is 30 lb per ton of weight.

Curve resistance is a function of the radius of curvature and gage of the track, wheelbase and diameter, speed, and load. It is often ignored because, in a properly installed system, it accounts for less than 1 lb per ton per degree of curve for that part of the train on the curve.

On a 1% grade, a ton (2,000 lb) must be raised 1 ft for every 100 ft that the train advances, making the grade resistance 20 lb per ton for each 1% of grade. For example, the pull required to overcome a grade of 3% for a 10-ton car is 600 lb (20 lb per ton for each 1% of grade × 3% grade × 10 tons).

Tractive effort is also required to accelerate or decelerate. Therefore, because force is equal to mass multiplied by acceleration, and an acceleration rate of 1 mphps is equivalent to 1.46 fpsps, approximately 100 lb per ton is required to accelerate 1 ton, as follows:

$$\text{Force} = F = ma = \frac{2{,}000 \text{ lb}}{(32.2 \text{ fpsps})(1.46 \text{ fpsps})} = 91.0 \text{ lb}$$

which is approximately 100 lb. Thus, in mine locomotive calculations, it is assumed that it takes 100 lb to accelerate each ton of the train's weight to achieve an acceleration rate of 1 mphps. Normally, mine locomotives accelerate between 0.1 and 0.2 mphps; therefore, the acceleration resistance is usually between 10 and 20 lb per ton. Naturally, when the trip is decelerating, it can be analyzed in a similar manner.

General Equations

The following equations relate the tractive effort that can be exerted by a locomotive to the forces that must be overcome.

Tractive Effort Required for Hauling on Grade Against Loads

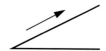

$$TE = W_L(F_L + 20g + 100a) + W_T(F_T + 20g + 100a) \qquad \text{(EQ 8.10)}$$

Braking conditions:

$$TE = W_L(-F_L - 20g + 100a) + W_T(-F_T - 20g + 100a) \qquad \text{(EQ 8.11)}$$

Tractive Effort Required for Hauling When Grades Are in Favor of Loads

$$TE = W_L(F_L - 20g + 100a) + W_T(F_T - 20g + 100a) \qquad \text{(EQ 8.12)}$$

Braking conditions:

$$TE = W_L(-F_L + 20g + 100a) + W_T(-F_T + 20g + 100a) \qquad \text{(EQ 8.13)}$$

where TE is the locomotive weight, in pounds, multiplied by adhesion; W_L is the locomotive weight in tons; W_T is the train weight in tons; F_L is the locomotive rolling resistance in pounds per ton; F_T is the train rolling resistance in pounds per ton; g is the percent grade; and a is the acceleration/deceleration rate in miles per hour per second.

LOCOMOTIVE MOTOR DUTY CYCLE

Occasionally, mining engineers must determine whether or not the motor capacity of the selected locomotive is sufficient to operate without overheating. To calculate this, a duty cycle for a round trip under operating conditions is determined; this, in turn, is related to the characteristic curves of the motors in the proposed locomotive. The procedure for the analysis is to determine the track profile and calculate the following values:

1. Tractive effort due to friction and grade resistances, from equations
2. Velocity and amperage from characteristic curves. (Note: If the curve's velocity exceeds the allowable velocity, on–off times must be determined.)
3. Time to travel the track-profile segment at the velocity determined in number 2

When all individual times and amperages are calculated for each segment, a special form of the rms (root mean square) method used for mine-hoist motors is applied:

$$I^2 = \frac{I_1^2 t_1 + I_2^2 t_2 + I_3^2 t_3 + \ldots}{\text{total cycle time}} \quad \text{(EQ 8.14)}$$

Taking the square root of this value yields the I_{rms} for each locomotive motor. This value (required amperage) is then compared with the allowable amperage based on the temperature curves.

Corrections may be required for high-altitude operation or where the ambient temperature varies from that assumed for the development of the curves. In these cases, the allowable amperage can be calculated using the following equation:

$$\ln(I) = \frac{[(\ln(H_T) - W)\ln(I_1) - \ln(I_2)]}{(Z - W)} + \ln I_2 \quad \text{(EQ 8.15)}$$

where:

H_T = total operation time for the motors in hours

$$W = \ln \frac{[(I_1^2 H_2)^{0.5}(T_P/T_c)]^2}{I_2^2}$$

$$Z = \ln \frac{[(I_1^2 H_1)^{0.5}(T_P/T_c)]^2}{I_1^2}$$

T_c = the rated allowable temperature rise in degrees Celsius

T_P = the corrected allowable temperature rise due to elevation or ambient temperature in degrees Celsius

H_1, I_1 = a point taken off of the manufacturer's time-to-rise temperature curve in hours, amp

H_2, I_2 = a second point taken from the curve in hours, amp

I = allowed rms current for the total operation time H_T in amp

To correct for elevation and ambient temperatures, these procedures must be followed:

1. The allowable temperature rise is applicable up to an elevation of 3,300 ft. Above this value, the allowable temperature rise is reduced 1% for every 330 ft, or:

$$\% \text{ reduction} = \frac{\text{elevation} - 3,300}{330} \quad \text{(EQ 8.16)}$$

2. When the ambient temperature exceeds 25°C, the allowable rise must be reduced by the difference between the ambient temperature and 25°C.

BELT CONVEYORS

Belt conveyors are high-capacity, reliable transportation systems for mined material. They represent the primary means of intermediate haulage in most of today's underground mines and have gained acceptance for main-line haulage. Developments in overland belt systems have also extended their influence to surface facilities. The major components of a belt-conveyor system are the belt, terminal pulleys, idlers (the cylindrical metal rollers that support the belt), support structure, takeup (tensioning) device, motor and drive assemblies, and numerous protection devices.

When designing belt-conveyor systems, mining engineers are concerned primarily with calculating the necessary belt tensions, takeup force (if necessary), idler spacings, and required motor horsepower. More detailed design information, such as belt width, speed, and construction, can be determined from manufacturers' handbooks, of which there are many. The following sections outline the design procedure for the calculations mentioned previously.

Belt Width and Speed

The acceptable operating ranges for the widths and speeds of conveyor belts are often interrelated; this relationship is due to the fact that an increase in belt speed can permit a reduction in belt width for any given mined material to be conveyed. To determine the minimum belt width required, information must be provided on the maximum size of the lumps to be conveyed and whether or not the lumps are uniform in size. For example, if run-of-mine coal (lumps and fines together) is to be conveyed at a rate of 1,200 tph, Table 8.3 can be used to indicate a minimum belt width of 36 in. when the maximum lump size is 12 in. Note that a 60-in. belt would be required if the lumps were uniformly sized at 12 in.

Once a minimum belt width (based on maximum lump size) has been determined, the speed of the belt can be established. Industry experience shows that the most economical selection usually means using maximum belt speeds. Using Table 8.4, the maximum belt speed in the case given is 650 fpm. To see if that speed is adequate, Table 8.5 should be consulted. If, as in this example, 35° troughing idlers are to be used and mined coal weighs 50 lb per cu ft, a 36-in.-wide belt can accommodate 159 tph of coal for each 100 fpm of belt speed. If the belt operates at 650 fpm, the maximum capacity is rather low: 1,033.5 tons. Thus, the next widest belt (42 in.), and its accompanying maximum speed (700 fpm), should be analyzed. In this instance, it is adequate (7 × 220 = 1,540 tph; 1,540 tph > 1,200 tph).

Belt Tension

A belt conveyor can be used to transport mined material not only along the horizontal but also along a change in grade (Figure 8.4). The power to move the belt is supplied by the motor through a drive pulley. The maximum tension developed in the belt is the sum of the following: (1) tension to overcome the friction of the empty belt and the conveyor components that come in contact with the empty belt, (2) tension to overcome the friction of the load, and (3) tension to raise or lower the load and belt through elevation changes.

TABLE 8.3 Maximum recommended lump size for various belt widths

Belt Width, in.	Lump Size, in.	
	If Uniform	If Mixed with 90% Fines
12	2	4
18	4	6
24	5	8
30	6	10
36	7	12
42	8	14
48	10	16
54	11	20
60 and over	12	24

Source: Goodyear Tire and Rubber Company 1975.

TABLE 8.4 Typical maximum belt speeds in feet per minute

Belt Width, in.	Grain or Other Free-flowing Material	Run of Mine Coal and Earth*	Hard Ores and Stone, Primary Crushed†
14	400	300	300
16	500	300	300
18	500	400	350
20	600	400	350
24	600	500	450
30	700	600	550
36	800	650	600
42	800	700	600
48	900	700	650
54	1,000	700	650
60	1,000	700	650
66	—	800	750
72	—	800	750

Source: Goodyear Tire and Rubber Company 1975.
Note: These speeds are intended as guides to general practice and are not absolute.
*Moderately abrasive materials.
†Very abrasive materials.

The empty conveyor friction force consists of the sum of the return-side friction force ($CQL/2$) and the carrying-side friction force when empty $\{CQ[(L/2) + L_o]\}$, where C is the idler friction factor, Q is the weight of the moving parts of the conveyor system expressed as pounds per foot of the conveyor's center-to-center distance, L is the horizontal projected length of the conveyor, and L_o is the friction equivalent. The idler friction factor depends on the type of idler, the structure, and the quality of the maintenance. The friction equivalent is used as a means of including the constant frictional losses of the terminal pulleys, which are independent of belt length. Table 8.6 provides values for C and L_o. Table 8.7 provides values for Q.

TABLE 8.5 Normal bulk material capacity of troughed conveyor belts in tons per hour for each 100 fpm of belt speed

Material	Idler Roll Angle, degrees	Material Density, per cu ft	Width, in.																		
			14	16	18	20	24	30	36	42	48	54	60	66	72	78	84	90	96	102	108	114	120
Most bulk materials Surcharge angle: 25° Edge distance of load: (0.055W + 0.9) in.	20	30	10	13	17	22	33	53	78	108	144	183	228	279	335	396	462	533	608	688	774	864	959
		50	16	22	28	36	55	88	130	180	240	305	380	465	557	661	770	887	1,013	1,147	1,289	1,440	1,599
		75	24	32	42	54	83	132	195	270	360	458	570	697	837	992	1,153	1,330	1,519	1,722	1,933	2,160	2,400
		100	32	43	56	72	110	176	260	360	480	610	760	930	1,115	1,321	1,539	1,774	2,026	2,294	2,579	2,880	3,198
		125	40	54	70	90	138	220	325	450	600	762	950	1,163	1,395	1,653	1,923	2,217	2,532	2,869	3,222	3,600	3,999
		150	48	65	84	108	165	264	390	540	720	915	1,140	1,395	1,672	1,982	2,309	2,661	3,039	3,441	3,868	4,320	4,797
	35	30	12	16	20	26	40	65	95	132	176	224	278	341	408	485	565	652	745	842	948	1,058	1,172
		50	19	27	34	44	67	108	159	220	293	373	464	568	680	809	943	1,086	1,240	1,404	1,580	1,763	1,958
		75	29	40	51	66	100	161	238	329	439	558	696	852	1,020	1,214	1,412	1,628	1,860	2,105	2,370	2,645	2,935
		100	39	53	68	88	134	215	317	439	585	745	928	1,135	1,360	1,618	1,885	2,172	2,480	2,808	3,160	3,526	3,915
		125	49	66	85	110	168	269	396	549	732	932	1,160	1,420	1,700	2,023	2,355	2,714	3,100	3,509	3,950	4,408	4,893
		150	59	80	102	132	201	322	476	660	878	1,118	1,392	1,703	2,040	2,427	2,828	3,258	3,720	4,212	4,740	5,289	5,873
	45	30	13	17	22	28	43	69	101	141	187	238	296	363	435	514	599	692	789	893	1,003	1,121	1,242
		50	21	28	37	47	72	115	169	234	312	397	494	605	725	857	999	1,151	1,314	1,488	1,672	1,868	2,073
		75	32	42	55	71	107	172	244	352	468	595	741	908	1,088	1,287	1,499	1,725	1,973	2,240	2,510	2,800	3,110
		100	42	56	73	94	143	229	338	468	624	793	988	1,210	1,450	1,715	1,998	2,302	2,628	2,976	3,344	3,735	4,146
		125	53	70	91	117	179	286	422	586	780	990	1,235	1,512	1,810	2,144	2,498	2,876	3,287	3,738	4,182	4,668	5,183
		150	63	84	110	141	214	344	507	702	936	1,190	1,482	1,815	2,175	2,572	2,997	3,453	3,942	4,464	5,016	5,603	6,219
Maximum recommended lump size*	Uniform size		2	3	4	4	5	6	7	8	10	11	12	12	12	12	12	12	12	12	12	12	12
	Mixed with fines		4	5	6	6	8	10	12	14	16	20	24	24	24	24	24	24	24	24	24	24	24

Note: Obtain capacities of their material densities and belt speeds by direct interpolation.
*Larger lumps can often be considered with special impact constructions and loading point designs.

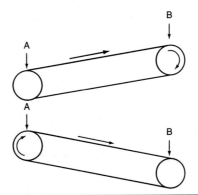

FIGURE 8.4 Conveyor belt arrangements
Source: Goodyear Tire and Rubber Company 1975.

The load friction force is equal to:

$$C(L + L_o)\frac{100T}{3S} \qquad \text{(EQ 8.17)}$$

where T is the peak capacity in tons per hour and S is the belt speed in feet per minute.

The force required to overcome the load on an incline is:

$$\pm\frac{100(T)(H)}{3S} \qquad \text{(EQ 8.18)}$$

where H is the elevation change in feet. This component of the tension equation takes on the negative sign when the discharge point is lower in elevation than the loading point.

The force required to overcome the belt weight on an incline is:

$$(B)(H) \qquad \text{(EQ 8.19)}$$

where B is the belt weight in pounds per foot of the conveyor's center-to-center distance.

The maximum calculated tension (T_1) represents the tight-side tension on one side of the drive pulley. To prevent slipping, a slack-side tension (T_2) is developed on the other side of the drive pulley. The difference between the two is called the effective tension (T_e) and is equal to:

$$T_1 - [(B)(H)] \qquad \text{(EQ 8.20)}$$

The minimum slack-side tension required becomes:

$$(T_e)(K) \qquad \text{(EQ 8.21)}$$

where K is the drive factor, which is a measure of the drive pulley's ability to transmit its torque without slippage. Values of K depend on the arc of contact between the belt and drive pulley(s), type of takeup, and whether the drive pulleys are bare or lagged. Table 8.8 provides values for K.

If the minimum slack-side tension is less than BH, no additional takeup tension force is required. Since any additional force will be divided evenly between the tight side

TABLE 8.6 Friction and length factors for conveyor belts

Class of Conveyor	Friction Factor (C)	Length Factor (L_o), ft
For conveyors with permanent or other well-aligned structures and with normal maintenance	0.022	200
For temporary, portable, or poorly aligned conveyors. Also for conveyors in extremely cold weather that either are subject to frequent stops and starts or are operating for extended periods at −40°F or below	0.03	150
For conveyors requiring restraint of the belt when loaded	0.012	475

Source: Goodyear Tire and Rubber Company 1975.

TABLE 8.7 B and Q for conveyor belts

Width, in.	Light-service Material to 50 lb per cu ft		Medium-service Material Over 50 But Not Exceeding 100 lb per cu ft		Heavy-service Material over 100 lb per cu ft	
	B	Q	B	Q	B	Q
14	1	7	2	13	3	19
16	2	8	3	14	4	21
18	3	9	4	16	5	23
20	4	10	5	18	6	25
24	5	14	6	21	7	29
30	6	19	7	28	8	38
36	7	26	9	38	11	52
42	9	33	11	50	14	66
48	12	40	15	60	18	82
54	14	50	18	71	22	97
60	17	62	21	85	27	115
66	20	75	24	103	32	135
72	22	88	28	121	36	155

Source: Goodyear Tire and Rubber Company 1975.

and slack side of the drive pulley, the takeup force must be twice as large as the difference between T_2 and BH:

$$\text{takeup tension force} = 2(T_2 - BH) \quad \text{(EQ 8.22)}$$

A belt must be capable of withstanding the maximum tension to be applied:

$$T_{max} = T_1 + 0.5 \text{ (takeup tension force)} \quad \text{(EQ 8.23)}$$

Idler Spacing

Table 8.9 lists typical spacing recommendations for conveyor carrying and return idlers.

TABLE 8.8 K values for conveyor belts

Arc of Contact, degrees	Manual Takeup		Automatic Takeup	
	Bare Pulley	Lagged Pulley	Bare Pulley	Lagged Pulley
150	1.20	1.00	0.84	0.67
180	0.97	0.80	0.64	0.50
190	0.91	0.75	0.59	0.46
200	0.85	0.71	0.54	0.42
210	0.80	0.66	0.50	0.38
220	0.75	0.62	0.46	0.35
230	0.72	0.59	0.43	0.33
240	0.68	0.56	0.40	0.30
270	0.58	0.49	0.32	0.24
300	0.51	0.43	0.26	0.19
330	0.46	0.40	0.22	0.16
360	0.42	0.36	0.18	0.13
390	0.39	0.33	0.15	0.11
420	0.36	0.31	0.13	0.09
450	0.33	0.29	0.11	0.07
480	0.31	0.27	0.09	0.06

Source: Goodyear Tire and Rubber Company 1975.

TABLE 8.9 Idler spacing for conveyor belts

Belt Width, in.	Carrying Idler Spacings			Return Idler Spacings, ft
	Material Weight, ft			
	Up to 50 lb	To 100 lb	100 lb or More	
14, 16, 18	5½	5	5	10
24, 30	5½	4	3½	10
36, 42	5	3½	3	10
48	4½	3½	2½	10
54	4½	3	2½	10
60	4½	3	2½	8 to 10
72	4½	3	2½	8 to 10

Source: Goodyear Tire and Rubber Company 1975.

Motor Horsepower

The required motor horsepower can be calculated using the following equation:

$$HP = \frac{ST_e}{33,000E} \quad \text{(EQ 8.24)}$$

where S is the belt speed in feet per minute, T_e is the effective tension in pounds, and E is the motor drive efficiency expressed as a decimal.

REFERENCES

Bethlehem Steel Co. 1958. *Mine and Industrial Trackwork for Safe Haulage.* Catalog 470. Bethlehem, PA: Bethlehem Steel Company.
Brauns, J.W., and D.H. Orr, Jr. 1968. Railroad. In *Surface Mining.* Edited by E.P. Pfleider. New York: AIME.
Goodyear Tire and Rubber Co. 1975. *Handbook of Conveyor and Elevator Belting.* Akron, OH: Goodyear Tire and Rubber Company.
Lewis, R.S., and G.B. Clark. 1964. *Elements of Mining.* New York: John Wiley and Sons, Inc.
Staley, W.W. 1949. *Mine Plant Design.* New York: McGraw-Hill, Inc.

PROBLEMS

Problem 8.1

Given the following information, determine the specifications for a switch to be used with 85-lb rail in an underground mine:

1. Cast frog used
2. Frog No. (N) = 5
3. Angle (F) = 11°25'16" = 11.421°
4. Length (L) = 120 in.
5. Gage (G) = 42 in.
6. Thickness of the frog point (B) = 0.5 in.
7. Thickness of the switch point (P) = 0.375 in.
8. Heel spread of switch (H) = 6.25 in.
9. Distance from the actual point of a frog to its toe (A) = 18.1875 in.

Problem 8.2

A 50-ton locomotive is to be used to haul coal from a very large loaded sidetrack 8,000 ft to the outside. Trips are to be designed to accelerate at a rate of 0.2 mphps up a 1% grade. If 10-ton capacity cars with a tare weight of 5 tons are used, how many locomotives will be required to haul a shift tonnage of 6,000 if the locomotives average 10 mph with 5-min turnaround time at each end (total of 10 min delay per trip)? Roller bearings are employed throughout and an adhesive factor of 25% is to be used. The operating time per shift is approximately 6 hr.

Problem 8.3

As the locomotive in Problem 8.2 was pulling up a 2% grade, the power went off when the locomotive was 1,000 ft from the top of the grade. When the power returned, the trip was accelerated to maximum design capacity on unsanded rails. How many miles per hour was the locomotive traveling when it reached the top of the hill and how long did this take?

Problem 8.4

As the locomotive in Problem 8.2 was descending a 1.5% grade at a speed of 10 mph, the motorman saw a fall of rock on the track 120 ft ahead. With sanding, he was able to apply an adhesive force of 0.30. Was he able to stop the trip in time and if so by how much?

Problem 8.5

Determine the size of locomotive required to pull a 300-ton train from a level siding at a maximum acceleration of 0.3 mphps if antifriction bearings with an F = 15 lb per ton are employed on all equipment. Use an adhesive factor of 30%.

Problem 8.6

Using the same size of locomotive and trip calculated in problem 8.5, determine the maximum grade against which this locomotive can pull the train without decelerating. Assume that the adhesive value is still 0.3.

Problem 8.7

Using the same size of locomotive and trip calculated in problem 8.5, determine the downgrade at which braking is impossible at any distance. Assume that braking will be attempted on sanded rails.

Problem 8.8

What motor horsepower must the locomotive used in problem 8.5 have if it is expected to operate at rated tractive effort (transmission efficiency: 95%, adhesive coefficient: 35%) at a speed of 8 mph?

Problem 8.9

What weight of rail should be used for main-line haulage if the mine uses 35-ton four-wheel haulage locomotives?

Problem 8.10

An 8-ton locomotive is to be used to haul nine 2-ton cars with a 4-ton load in each car. Using the characteristic curves shown, determine the rms amperage and whether or not the locomotive is suitably designed to run continuously for 6 hr per shift. The following information also applies:

1. Track profile (from panel to shaft bottom):
    a. 600 ft of level track
    b. 2,000 ft at −0.5%
2. Maximum speed: 8 mph
3. Roller bearings used throughout
4. Two motors on the locomotive
5. Loading requirements: 2 min at 60 amp

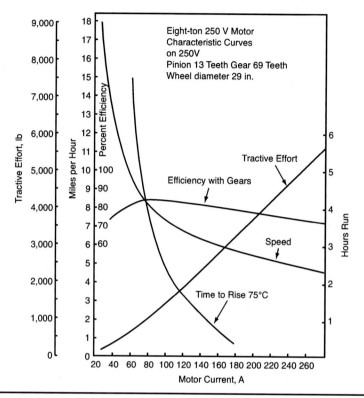

Illustrates Problem 8.10

6. Unloading requirements: 2 min at 40 amp
7. Elevation of the haulage entry: 5,000 ft
8. Temperature of the haulage entry: 30°C
9. Shift tonnage to be transported: 500 tons

Problem 8.11

Design a 42-in. conveyor belt to haul coal (55 lb per loose cubic ft) 3,000 ft at a level grade in an underground mine. The peak capacity should be 500 tph, and the belt speed is projected to be 600 fpm. The drive has an automatic takeup, lagged pulley, and a 240° arc of contact; the motor drive efficiency is 0.85.

Problem 8.12

A 60-in. wide conveyor belt is to be designed for coal haulage up an 1,857-ft-long 18° slope. The peak capacity is estimated to be 1,000 tph, and the belt speed will be set at 700 fpm. The belt drive will have an automatic takeup, lagged pulley, and a 180° arc of contact. If motor drive efficiency is assumed to be 90%, determine the various belt tensions and the required motor horsepower. Assume that the coal weighs 60 lb per loose cubic foot.

PROBLEM SOLUTIONS

Solution 8.1

$$\text{Switch angle (S)} = \arcsin \frac{H - P}{L}$$

$$= \arcsin \frac{6.25 - 0.375}{120}$$

$$= 2.806° = 2°48'22"$$

$$\text{Theoretical toe distance (E)} = A - (B)(N)$$

$$= 18.1875 - (0.5)(5)$$

$$= 15.6875$$

$$15.6875 + 10 \text{(due to cast frog)} = 25.6875 \text{ in.}$$

$$\text{Theoretical lead (X)} = (L \cos S) + \frac{G - H - E \sin F}{\tan 0.5(F + S)} + E \cos F$$

$$= (120)(0.9988) + \frac{42 - 6.25 - (25.6875(25.6875))(0.198)}{0.125} + 25.6875(0.98)$$

$$= 119.856 + 245.311 + 25.174 = 390.34 \text{ in.}$$

$$\text{Actual lead (Y)} = X + BN = 390.34 + (0.5)(5) = 392.84 \text{ in.}$$

$$\text{Radius (R)} = \frac{G - H - E \sin F}{2 \sin 0.5(F + S) \sin 0.5(F - S)}$$

$$= 42 - 6.25 - \frac{(25.6875)(0.198)}{2(0.124)(0.075)}$$

$$= 1{,}649 \text{ in.} = 137.4 \text{ ft}$$

$$\text{Length of straight closure rail (D)} = X - L - E + 10$$

$$= 390.34 - 120 - 25.6875 + 10$$

$$= 254.65 \text{ in.} = 21.22 \text{ ft}$$

$$\text{Length of curved closure rail (C)} = R (F - S) + 10$$

$$= (1{,}659)\left(\frac{11.421 - 2.806}{360}\right) 2\pi + 10$$

$$= 257.9 \text{ in.} = 21.50 \text{ ft}$$

$$\text{Middle ordinate (MO)} = R - [R \cos 0.5 (F - S)]$$

$$= 1{,}649 - [1{,}649 (0.9972)] = 4.62 \text{ in.}$$

Solution 8.2

$$TE \text{ (lb)} = W_L (F_L + Gg + Aa) + W_T (F_T + Gg + Aa)$$

$$50 (2{,}000)(0.25) = 50 (20 + 20 + 20) + W_T (20 + 20 + 20)$$

$$W_T = (25{,}000 - 3{,}000)/60 = 366 \text{ tons}$$

Number of cars = $\dfrac{366}{(10 + 5)}$ = 24.4 or 24 cars

$(24)(10) = 240$ tons per trip

$$(10 \text{ mph})(5{,}280 \text{ ft per mile})(1 \text{ hr per 60 min}) = 880 \text{ fpm}$$

$$\frac{16{,}000}{880} = 18.2 \text{ min (round trip)}$$

In 6 hr:

$$\frac{(6)(60)}{(18.2 + 10)} = \frac{360}{28.2} = 12.76 \text{ trips per shift}$$

It will be assumed that over the long run, because the operating time per shift is approximately 6 hr, 12.76 trips per shift will be averaged. This number is used for calculations rather than rounding down to 12 trips per shift or up to 13 trips per shift. It follows that the tons per shift will be 3,062[12.76(240)]. Two locomotives are required for 6,000 tons per shift.

Solution 8.3

$$TE \text{ (lb)} = W_L (F_L + Gg + Aa) + W_T (F_T + Gg + Aa)$$

$$50 (2{,}000)(0.3) = 50 (20 + 40 + 100a) + 360 (20 + 40 + 100a)$$

$$a = \frac{5{,}400}{41{,}000} = 0.1317 \text{ mphps}$$

$$(0.1317 \text{ mphps})(1.46) = 0.1923 \text{ fpsps}$$

$$S = v_o t + 0.5at^2 = 0 + 0.5at^2$$

$$t = \left(\frac{2s}{a}\right)^{0.5} = \left(\frac{(2)(1{,}000)}{0.1923}\right)^{0.5} = 101.98 \text{ sec}$$

$$v = v_o + at = 0 + at = 0.1923 \,(101.98 \text{ sec}) = 19.61 \text{ fps}$$

$$(19.61 \text{ fps})(1 \text{ mile per } 5{,}280 \text{ ft})(3{,}600 \text{ sec per hr}) = 13.37 \text{ mph}$$

Solution 8.4

$$\text{TE (lb)} = W_L(-F_L + Gg + Aa) + W_T(-F_T + Gg + Aa)$$

$$(50)(2{,}000)(0.3) = 50(-20 + 30 + 100a) + 360(-20 + 30 + 100a)$$

$$a = \frac{25{,}900}{41{,}000} = 0.632 \text{ mphps}$$

$$(0.632 \text{ mphps})(1.46) = 0.9227 \text{ fpsps}$$

$$v = 0 = v_o + at = 14.67 \text{ fps} - 0.9227 \text{ fpsps}(t)$$

$$t = \frac{14.67}{0.9227} = 15.90 \text{ sec}$$

$$S = v_o t + 0.5at^2$$

$$= 14.67 \text{ fps}(15.9 \text{ sec}) + (0.5)(-0.9227 \text{ fpsps})(15.9 \text{ sec})^2 = 116.63 \text{ ft}$$

Yes, he was able to stop 3.36 ft short of the rock.

Solution 8.5

$$\text{TE} = W_L(F_L + 20g + 100a) + W_T(F_T + 20g + 100a)$$

$$W_L(2{,}000)(0.3) = W_L(15 + 0 + 30) + 300(15 + 0 + 30)$$

$$W_L = \frac{13{,}500}{(600 - 45)} = 24.32 \text{ tons}$$

Solution 8.6

$$\text{TE} = (W_L + W_T)(F + 20g)$$

$$(24.32)(2{,}000)(0.3) = (324.32)(15 + 20g)$$

$$g = 1.5\%$$

Solution 8.7

$$\text{TE} = (W_L + W_T)(-F + 20g + 100a)$$

Since braking is impossible, the value of a in the preceding equation is 0.0.

$$(24.32)(2{,}000)(0.3) = (324.32)(-15 + 20g)$$

$$\frac{14{,}592}{324.32} = -15 + 20g$$

Therefore, g is greater than 3.0%.

Solution 8.8

$$\text{TE} = 24.32(2{,}000)(0.35) = 17{,}024$$

$$\text{HP} = \frac{(\text{TE})(\text{MPH})}{375(E_T)} = \frac{(17{,}024)(8)}{375(0.95)} = 382 \text{ hp}$$

Solution 8.9

Using Table 8.1, 85-lb rail appears to be adequate.

Solution 8.10

To solve a problem of this nature, take the total train weight for each track-section profile, consider the effects of grade and friction resistance on the locomotive's tractive effort requirements, and calculate the tractive effort required per motor for the locomotive.

Section 1 (Loaded, Level Track):

$$TE = (W_L + W_T)(F + 20g)$$

$$= (8 + 54)(20 + 0) = 1{,}240 \text{ lb}$$

Since the locomotive has two motors, the TE per motor is 620 lb. From the characteristic curves (-----) shown in the illustration, I = 55 amp, speed = 10 mph. However, the maximum allowable speed is 8 mph. In this situation, the locomotive would commonly be operated on–off, on–off, etc., to maintain but not exceed 8 mph throughout this position of the track profile. To determine the time that the motors are on and off, find the time required to travel the distance at the maximum allowed speed and determine the time it would take at the curve speed; the difference between the two times represents the time at zero current.

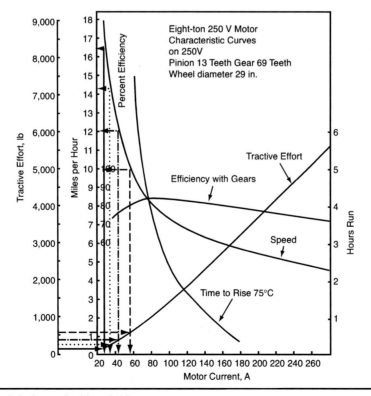

Illustrates Solution to Problem 8.10

At 10.0 mph:

$$t = (600 \text{ ft})/[(88 \text{ fpm/mph})(10 \text{ mph})] = 0.68 \text{ min}$$

At 8.0 mph:

$$t = (600 \text{ ft})/[(88 \text{ fpm/mph})(8 \text{ mph})] = 0.85 \text{ min}$$

Thus, the zero I time = 0.85 − 0.68 = 0.17 min

The duty cycle for this section is:

0.68 min at 55 amp
0.17 min at 0 amp

Section 2 (Loaded, − 0.5% Grade):

$$TE = (W_L + W_T)(F - 20 g)$$

$$= (62)(20 - 10) = 620 \text{ lb}$$

Since the locomotive has two motors, the TE per motor is 310 lb. From the characteristic curves ($\cdots\cdots\cdots$) shown earlier, I = 35 amp, speed = 14.3 mph. Again, the maximum allowable speed is exceeded.

At 14.3 mph:

$$t = (2000 \text{ ft})/[(88 \text{ fpm/mph})(14.3 \text{ mph})] = 1.59 \text{ min}$$

At 8.0 mph:

$$t = (2000 \text{ ft})/[(88 \text{ fpm/mph})(8.0 \text{ mph})] = 2.84 \text{ min}$$

Thus, the zero I time = 2.84 − 1.59 = 1.25 min

The duty cycle for this section is:

1.59 min at 35 amp
1.25 min at 0 amp

Section 2 (Unloaded, +0.5% Grade):

$$TE = (W_L + W_T)(F + 20g)$$

$$= (8 + 18)(20 + 10) = 780 \text{ lb}$$

Since the locomotive has two motors, the TE per motor is 390 lb. From the characteristic curves (— · — · —) shown earlier, I = 44 amp, speed = 12.1 mph. Again, the maximum allowable speed is exceeded.

At 12.1 mph:

$$t = \frac{2{,}000 \text{ ft}}{(88 \text{ fpm/mph})(12.1 \text{ mph})} = 1.88 \text{ min}$$

At 8 mph:

$$t = \frac{2{,}000 \text{ ft}}{(88 \text{ fpm/mph})(8.0 \text{ mph})} = 2.84 \text{ min}$$

Thus, the zero I time = 2.84 − 1.88 = 0.96 min

The duty cycle for this section is:

    1.88 min at 44 amp
    0.96 min at 0 amp

Section 1 (Unloaded, Level Track):

$$TE = (W_L + W_T)(F + 20g) = (26)(20 + 0) = 520 \text{ lb}$$

Since the locomotive has two motors, the TE per motor is 260 lb. From the characteristic curves (———) shown earlier, I = 30 amp, speed = 16.6 mph. Again, the maximum allowable speed is exceeded.

At 16.6 mph:

$$t = \frac{600 \text{ ft}}{(88 \text{ fpm/mph})(16.6 \text{ mph})} = 0.41 \text{ min}$$

At 8 mph:

$$t = \frac{600 \text{ ft}}{(88 \text{ fpm/mph})(8.0 \text{ mph})} = 0.85 \text{ min}$$

Thus, the zero I time = 0.85 − 0.41 = 0.44 min

The duty cycle for this section is:

    0.41 min at 30 amp
    0.44 min at 0 amp

Total Trip Time:

Round-trip time = running time + loading time + unloading time

    = 0.68 + 0.17 + 1.59 + 1.25 + 1.88 + 0.96 + 0.41 + 0.44 + 2.0 + 2.0

    = 11.38 min

Trips per shift = shift tonnage/tons per trip = 500/36 = 13.89 ≈ 14 trips

Time per trip (allowable) = [(6 hr)(60 min per hr)]/14 trips = 25.7 min

This allows 14.33 min of delay time per trip (25.5 − 11.38 min). Zero current is drawn during this delay period.

Therefore, the rms current needed is:

$$\left(\frac{\text{Total } I^2 t_e}{\text{Total } t_e}\right)^{0.5} = 40.2 \text{ amp}$$

To find the allowable rms current, the following formula is used:

$$\ln(I) = \frac{(\ln H_T - W)(\ln I_1 - \ln I_2)}{Z - W} + \ln I_2$$

Using the characteristic curves shown previously, the following two points (denoted by Xs) will be taken off of the Time to Rise 75°C Curve:

$$H_1, I_1 : 6 \text{ hr}, 70 \text{ amp}$$

$$H_2, I_2 : 2 \text{ hr}, 110 \text{ amp}$$

Corrections must now be made for elevation and ambient temperatures:

$$\text{percent decrease due to elevation} = \frac{5{,}000 - 3{,}300}{330} = 5.15\%$$

and the maximum allowable motor temperature rise = 75 − 75 (0.0515) − 5 = 66.1°C

W and Z become:

$$W = \frac{[\ln(12{,}100 \times 2)^{0.5}(66.1/75)]^2}{12{,}100} = 0.44$$

$$Z = \frac{[\ln(4{,}900 \times 6)^{0.5}(66.1/75)]^2}{4{,}900} = 1.54$$

and:

$$\ln(I) = \frac{[\ln(6) - 0.44][\ln(70 - \ln(110))]}{1.54 - 0.44 + \ln(110)}$$

$$= \frac{(1.79 - 0.44)(4.25 - 4.70)}{1.10} + 4.7$$

$$= 4.15$$

Thus, $I_{\text{allowable}}$ = 63.5 amp and the locomotive/motor combination will work (40.2 amp < 63.5 amp).

Cycle-Time Tabulation:

Cycle	Section	TE per Motor	Actual Time, min	Effective Time (t_e), min	I	$I^2 t_e$
Loaded	1	620	0.68	0.68	55	2,057
			0.17	0.17	0	0
	2	310	1.59	1.59	35	1,948
			1.25	1.25	0	0
Unloaded	2	390	1.88	1.88	44	3,640
			0.96	0.96	0	0
	1	260	0.41	0.41	30	369
			0.44	0.44	0	0
Delay time			14.33	0	0	0
Loading time			2.00	2.00	60	7,200
Unloading time			2.00	2.00	40	3,200
Totals:				11.38		18,414

Solution 8.11

Empty conveyor friction force:

$$\frac{CQL}{2} = \frac{(0.022)(50)(3,000)}{2} = 1,650 \text{ lb}$$

$$CQ\left(\frac{L}{2} + L_o\right) = (0.022)(50)\left(\frac{3,000}{2}\right) + 200 = 1,870 \text{ lb}$$

$$\text{Total:} = 3,520 \text{ lb}$$

Load friction force:

$$C(L + L_o)\left(\frac{100T}{3S}\right) = (0.022)(3,000 + 200)\frac{(100)(500)}{(3)(600)} = 1,956 \text{ lb}$$

$$T_1 = 3,520 + 1,956 = 5,476 \text{ lb}$$

$$T_1 - T_2 = T_e = T_1 - BH$$

$$T_e = T_1 - 0.0 = 5,476 \text{ lb}$$

Thus, the minimum slack-side tension = T_e (K)

$$= 5,476(0.3)$$
$$= 1,643 \text{ lb}$$

$$\text{Takeup tension force} = 2(T_2 - BH)$$
$$= 2(1{,}643 - 0)$$
$$= 3{,}286 \text{ lb}$$
$$\text{Maximum applied tension} = T_1 + 0.5(\text{takeup tension force})$$
$$= 5{,}476 + (3{,}286/2) = 7{,}119 \text{ lb}$$

Idler spacing: From Table 8.9, the carrying-idler spacing is approximately 5 ft, while the return idlers can be spaced on 10-ft centers.

Motor horsepower:

$$HP = ST_e/33{,}000 \, E$$
$$= 600 \, (5{,}476)/[33{,}000 \, (0.85)]$$
$$= 117 \text{ hp}$$

Typically, two 75-hp motors would be placed in tandem in this situation.

Solution 8.12

Empty conveyor friction force:

$$\frac{CQL}{2} = \frac{(0.022)(85)(1{,}766)}{2} = 1{,}651 \text{ lb}$$

where $L = 1{,}857 \, (\cos 18°)$

$$CQ\left[\left(\frac{L}{2}\right) + L_o\right] = (0.022)(85)\left(\frac{1{,}766}{2} + 200\right) = \underline{2{,}025 \text{ lb}}$$

$$\text{Total} = 3{,}676 \text{ lb}$$

Load friction force:

$$C(L + L_o)\left(\frac{100T}{3S}\right) = (0.022)(1{,}766 + 200)\frac{100(1{,}000)}{3(700)} = 2{,}060 \text{ lb}$$

Inclined load force:

$$\frac{+100TH}{3S} = \frac{+100(1{,}000)(574)}{3(700)} = 27{,}333 \text{ lb}$$

where $H = 1{,}857(\sin 18°) = 574$ ft

Inclined belt force:

$$BH = 21(574) = 12{,}054 \text{ lb}$$
$$T_1 = 3{,}676 + 2{,}060 + 27{,}333 + 12{,}054 = 45{,}123 \text{ lb}$$
$$T_e = T_1 - T_2 = T_1 - BH = 45{,}123 - 12{,}054 = 33{,}069 \text{ lb}$$

Minimum slack-side tension:

$$T_2 = T_e(K) = 33,069(0.50) = 16,535 \text{ lb}$$

Because the minimum slack-side tension is greater than BH (12,054 lb), an additional takeup force is required:

$$\text{Takeup force required} = 2(16,535 - 12,054) = 8,962 \text{ lb}$$

$$T_{max} = T_1 + 0.5(\text{takeup force required}) = 45,123 + 0.5(8,962) = 49,604 \text{ lb}$$

$$\text{HP} = \frac{ST_e}{33,000}(E) = \frac{(700)(33,069)}{(33,000)(0.9)} = 779 \text{ hp}$$

An 800-hp motor would probably be selected.

CHAPTER 9

Rubber-Tired Haulage Systems

Face haulage in underground mines and pit haulage in surface mines are the principal domains of rubber-tired haulage vehicles. In underground mines, cable-reeled, battery-powered, or diesel-powered haulage vehicles are used; trucks are primarily used in surface mines.

Although many factors are considered when selecting rubber-tired haulage vehicles, such as clearance and tire life, the most important factor from a mine-planning standpoint is the production capability of the vehicle in the mine environment. This chapter addresses this topic.

UNDERGROUND FACE-HAULAGE VEHICLES

The selection of underground face-haulage vehicles is, in some respects, much simpler than that of the other haulage systems previously discussed. It is simpler because these vehicles, which transfer the mined material from the face to the intermediate haulage system, are manufactured in various sizes to cope with the confinement of underground-opening widths and heights. However, because these vehicles operate intermittently in conjunction with a constantly advancing face, the system's lack of permanence and overall dynamic nature create a more difficult situation for predicting productivity. The emphasis in this section is on modeling of productivity.

Three fundamental steps must be taken to analyze the productivity potential of any face-haulage system:

1. Recognize the geologic (seam thickness, etc.) and mining (opening widths, etc.) constraints that affect the velocity/clearance aspects of the haulage equipment and select equipment from manufacturers' specifications with these constraints in mind.
2. Secure the elemental times of the face-haulage system.
3. Incorporate the elemental times into an acceptable mine plan.

By recognizing the geologic and mining constraints that affect the system layout, parameters, such as vehicle payload, tonnage to be hauled per cut, and haulage distances, can easily be determined.

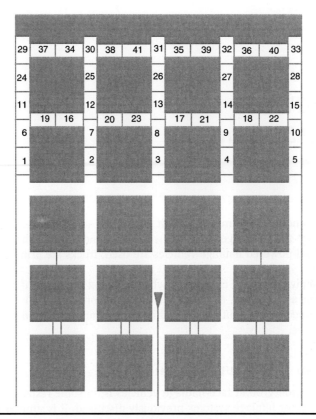

FIGURE 9.1 Sample cut sequence in an underground coal mine

Elemental times can be developed in two ways:

1. In an existing operation, a time study of the dynamic system can be conducted by placing three time-study engineers at the following static locations: (1) near the loading machine, (2) near the change point (CP) outby the face, and (3) near the dump. By synchronizing their watches and recording the actions of the haulage vehicles at these three locations, their time studies can be reduced to provide elemental times, such as vehicle tram rates (empty or loaded), vehicle discharge times, and loading rates.
2. In the absence of an existing operation, average values for the elemental times can be provided from other mines operating in similar conditions or by the manufacturers themselves.

Once the mining and geologic constraints are understood and the elemental times are secured, a typical layout, such as the one shown in Figure 9.1, can be analyzed. To do this, a cycle-time analysis is usually performed by either computer simulation or mathematical modeling. In this text, only mathematical modeling is considered. A spreadsheet incorporating the necessary equations can easily be constructed.

The mathematical model that is most frequently used for face-haulage studies is:

$$LCT = LT + COT + WSC + MISC + TT \qquad \textbf{(EQ 9.1)}$$

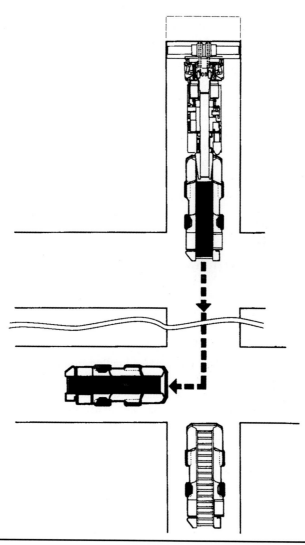

FIGURE 9.2 Haulage vehicles at the intersection containing the change point

where LCT is the loader cycle time for the cut in minutes; LT is the time that the loading machine actually spends loading the cut in minutes; COT is the time, in minutes, that the haulage vehicles tram inby the change point, which is located in the outby intersection where the empty haulage vehicle(s) waits for the loaded vehicle to clear the haulage path; and WSC is the time, in minutes, that the loader spends waiting for an empty face-haulage vehicle to arrive at the intersection containing the change point once the loaded vehicle clears the route inby. Figure 9.2 shows the situation when WSC equals zero; the empty vehicle has returned in time to advance inby the change-out point along a clear roadway.

MISC is the time, in minutes, resulting from a delay of the cycle to conduct miscellaneous operations, such as a methane check or advancement of ventilation tubing, and tramming the loader to the face of the next cut when an empty haulage vehicle is available at the loader.

TT is the time, in minutes, required to tram the loading machine from the completed cut to the face of the next cut. When a single-crew super section—a concept where two continuous miners are in the same section but only one is mining at any given time—is used, TT becomes the miner operator's time to walk from the recently completed cut to the continuous miner, which is set up to mine the next cut.

It is ironic that the haulage cycle-time analysis has the loader as its reference point; this merely proves the interrelationship between the loading and hauling operations. Three of the five components on the right-hand side of the equation can be further defined with the aid of the elemental times, rates, and capacities:

$$LT = \frac{T}{LR} \qquad (EQ\ 9.2)$$

$$COT = (N-1)\left(\frac{2COD}{SPD}\right) \qquad (EQ\ 9.3)$$

$$WSC = \frac{(N-1)}{2} * \{[(2\ HD/SPD) + DT] - \{[(NO-1)(CAP/LR)] + [2\ (COD/SPD)]\}\} \qquad (EQ\ 9.4)$$

where T is the tons in the cut; LR is the average loading rate in tons per minute; CAP is the average shuttle-car payload in tons; N, which is (T/CAP)*, is the number of shuttle-car loads in the cut [( )* means raise the calculated value within the parentheses to the next highest integer and ( )* means lower the calculated value within the parentheses to the next lowest integer]; COD is the one-way distance, in feet, traveled between the loader and the intersection containing the change point; SPD is the average shuttle-car speed in feet per minute; HD is the one-way distance traveled between the dump and the intersection containing the change point in feet; DT is the average discharge time of the haulage vehicle in minutes; and NO is the number of haulage vehicles in use.

An effective way to ensure that the COD and HD are measured correctly is to place a "dot" on the inby end of the haulage vehicles. The distance traveled by the "dot" from the loader to the point where it clears the intersection containing the change point (shown in Figure 9.2) is the COD. HD is measured from the point where the dot clears the change-point intersection to the dump. Since the "dot" is placed on the inby end of the vehicle, the vehicle length is automatically considered.

One technique that has been used to accommodate two shuttle cars sharing the same haulage path is called "switch-in/switch-out." In this situation, the empty standard shuttle car will "switch-in" into a left-side crosscut until the loaded opposite-standard shuttle car clears the intersection containing the change point. In this way, the loaded vehicle will not run over the cable of the empty vehicle. When clear, the standard shuttle car will "switch-out" or back out into the entry to proceed inby toward the awaiting loading machine or continuous miner. Obviously, when the empty opposite-standard shuttle car approaches the change point, it will "switch-in" and "switch-out" of the right-side crosscut.

Operationally, battery-powered or diesel-powered haulage vehicles are not constrained, as are cable-reel shuttle cars, because of the absence of the power cables. As a result, these untethered vehicles follow a cycle in which all loaded vehicles follow a

TABLE 9.1 Industry-accepted standards of rolling resistance factors (20 lb per ton = 1%)

	lb per ton	(kg/t)
A hard, smooth, stabilized, surfaced roadway without penetration under load; watered; maintained	40	(20)
A firm, smooth, rolling roadway with dirt or light surfacing; flexing slightly under load or undulating; maintained fairly regularly; watered	65	(35)
Snow, packed	50	(25)
Snow, loose	90	(45)
A dirt roadway; rutted; flexing under load; little if any maintenance; no water; 1 in. (25 mm) or 2 in. (50 mm) tire penetration	100	(50)
Rutted dirt roadway; soft under travel; no maintenance; no stabilization; 4 in. (100 mm) to 6 in. (150 mm) tire penetration	150	(75)
Loose sand or gravel	200	(100)
Soft, muddy, rutted roadway; no maintenance	200 to 400	(100 to 200)

Source: Caterpillar 1975.

common path to the dump and all empty vehicles follow a common path back to the loader. Therefore, it is much easier to add more vehicles to the haulage plan. However, two additional times must be added: a turn time so that the empty hopper can be pointed toward the loader and a turn time so that the loaded hopper can be pointed toward the dump. Since these two times occur outby the change point, they can be easily accommodated by the equations. To do this, merely add both times to the discharge time (DT) per vehicle.

When these equations are applied to a mining plan, the time required to load out a particular cut can be determined. Slight alterations in the mining plan (cut sequence, pillar dimensions, etc.) can be analyzed in the same way to evaluate their effect on overall productivity.

SURFACE HAULAGE VEHICLES

Trucks are the most widely used surface haulage vehicles. Although the selection procedure for haulage trucks is similar to that used to select other forms of wheeled mine-haulage vehicles, there are some differences that must be mentioned at the outset.

Trucks are rated in gross vehicle weight (GVW), which is equal to the weight of the empty vehicle plus the weight of the load. For a properly maintained truck, rolling resistance factors can be tabulated as shown in Table 9.1. Grade resistance (assistance) is equal to the GVW multiplied by the percent grade multiplied by 20 lb per ton, identical to the procedure followed for mine locomotives. Thus, the total resistance can be expressed as the sum of the rolling resistance and the grade resistance. This total resistance is referred to as effective grade; when multiplied by GVW, this is identical to the power required to move the vehicle.

Once the required power is determined, the amount of power available to move the truck is considered; the two factors that govern this are horsepower and speed. Trucks have transmissions, which provide speed and pull combinations that are designed to meet the requirements dictated by various situations. For example, assume that a certain off-highway truck has the characteristics shown in Figure 9.3, with a GVW of 150,000 lb and

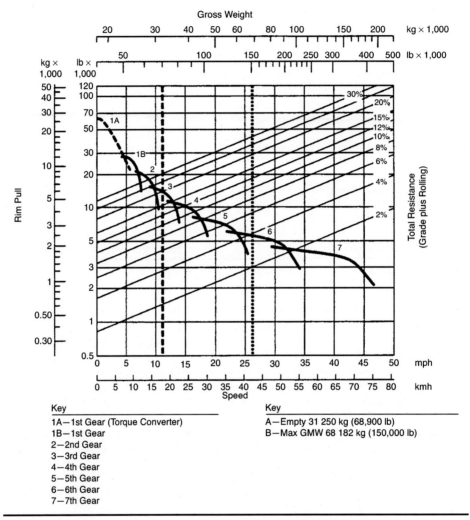

FIGURE 9.3 Typical rim-pull-speed-gradability curves for an off-highway truck
Source: Caterpillar 1997.

a total resistance of 6%. To determine the available power, find the 6% total resistance line and locate its point of intersection with a line dropped down from 150,000 lb of GVW. Moving to the left from this point, the rim pull (8,000 lb) can be read directly from the vertical axis. Notice that in moving to the rim-pull axis, a curve listed as fourth gear is also intersected. Dropping down from this point to the horizontal axis indicates that the power required is available when the truck is in fourth gear at a speed of 17 mph.

So far, the adhesion of the haulage truck has been considered to be 100% of the weight on the drive wheels. However, similar to the situation with locomotive haulage, this is never the case. Table 9.2 shows the approximate coefficients of traction for rubber-tired haulage trucks. Thus, if a truck has 80,000 lb on its rear wheels when loaded at its rated capacity on a concrete roadbed, its usable power would be 72,000 lb [80,000 lb (0.90)].

TABLE 9.2 Approximate coefficient of traction factors

	Rubber Tires	Tracks
Concrete	0.90	0.45
Clay loam, dry	0.55	0.90
Clay loam, wet	0.45	0.70
Rutted clay loam	0.40	0.70
Dry sand	0.20	0.30
Wet sand	0.40	0.50
Quarry pit	0.65	0.55
Gravel road (loose, not hard)	0.36	0.50
Packed snow	0.20	0.25
Ice	0.12	0.12
Semi-skeleton shoes	—	0.27
Firm earth	0.55	0.90
Loose earth	0.45	0.60
Coal, stockpiled	0.45	0.60

Source: Caterpillar 1997.

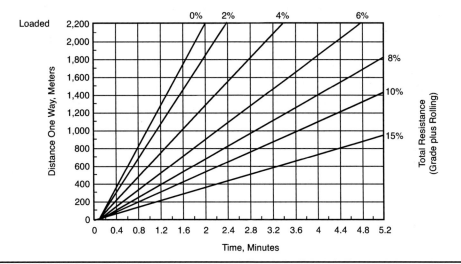

FIGURE 9.4 Travel time (loaded) for a typical off-highway truck
Source: Caterpillar 1997.

Performance Calculation

To determine the performance of a haulage truck, the following four items must be considered:

1. Haulage capacity in tons carried
2. Cycle time: The sum of loading, hauling, dumping, and return times for the truck (typical estimates for the travel times (loaded and empty) for a 138,000-lb GVW haulage truck are provided in Figures 9.4 and 9.5)

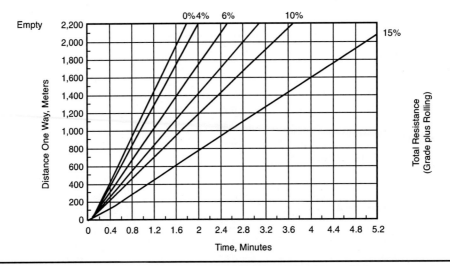

FIGURE 9.5 Travel time (empty) for a typical off-highway truck
Source: Caterpillar 1997.

3. Hourly production: the product of the cycles per hour, the load per cycle, and the job efficiency factor
4. Correction factors based on job conditions

When operating on steep downgrades, trucks may have to limit their operating speeds for safety. The maximum safe sustainable speed can be obtained by using performance curves, such as the ones shown in Figures 9.6 to 9.10, which relate to a 150,000-lb GVW truck. To determine the maximum safe speed, read from gross weight down to the percent effective grade. From that point, read horizontally to the curve with the highest obtainable speed range, then down to the maximum descent speed that the brakes can safely handle without exceeding their cooling capacity. Once the maximum safe speed is determined, the travel time for the segment can be determined using the following equation:

Time (in minutes) = (distance in feet)/[(88)(speed in miles per hour)] **(EQ 9.5)**

REFERENCES

Bishop, T.S. 1968. Trucks. In *Surface Mining*. Edited by E.P. Pfleider. New York: AIME.
Caterpillar Tractor Co. 1997. *Caterpillar Performance Handbook*. 28 ed. Peoria, IL: Caterpillar Tractor Co.
Caterpillar Tractor Co. 1975. *Fundamentals of Earthmoving*. Peoria, IL: Caterpillar Tractor Company.
Prelaz, L.J., et al. 1964. *Optimization of Underground Mining*. Report for Office of Coal Research. Washington, DC: USDI.

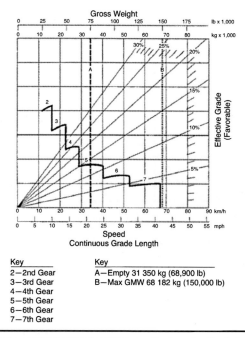

FIGURE 9.6 Typical brake performance (continuous grade retarding) for an off-highway truck
Source: Caterpillar 1997.

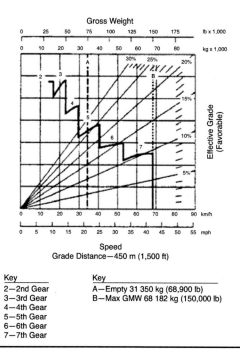

FIGURE 9.7 Typical brake performance (1,500 ft) for an off-highway truck
Source: Caterpillar 1997.

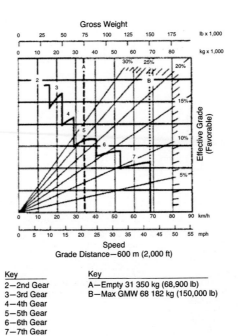

FIGURE 9.8 Typical brake performance (2,000 ft) for an off-highway truck
Source: Caterpillar 1997.

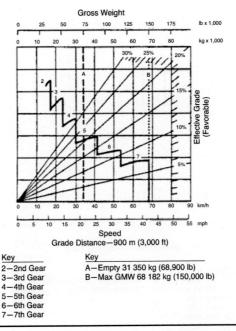

FIGURE 9.9 Typical brake performance (3,000 ft) for an off-highway truck
Source: Caterpillar 1997.

RUBBER-TIRED HAULAGE SYSTEMS | 251

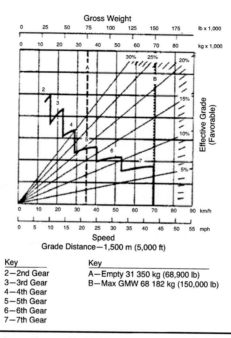

Key
2—2nd Gear
3—3rd Gear
4—4th Gear
5—5th Gear
6—6th Gear
7—7th Gear

Key
A—Empty 31 350 kg (68,900 lb)
B—Max GMW 68 182 kg (150,000 lb)

FIGURE 9.10 Typical brake performance (5,000 ft) for an off-highway truck
Source: Caterpillar 1997.

PROBLEMS

Problem 9.1

Shown next is a portion of a section layout in an underground coal mine. The cut to be mined is in the belt entry. Because of this, a "switch-in/switch-out" haulage system is to be tried, with the two cable-reel shuttle cars switching in the first crosscut outby the face. The face is located 70 ft inby the change point, and the cut will be driven 20 ft deep. For tram-distance calculations, measure from the midpoint of the cut. Based on the pertinent information shown below, determine the cycle time for the cut. List operating times for each shuttle car (one column for car A, which switches into the left crosscut, and one column for car B, which switches into the right crosscut). Check your answer with the mathematical model for face haulage operations. Start the cycle when the first ton of coal is loaded into the first shuttle car and end the cycle when the last ton is discharged at the dump. Neglect the time it takes to tram the continuous miner to the next cut.
Pertinent information:

   Seam thickness = 6 ft
   Entry and crosscut widths = 20 ft
   Pillar size = 60 ft × 60 ft
   Shuttle-car tram rate = 300 fpm
   Shuttle-car payload = 8 tons
   Continuous miner = 25 ft long

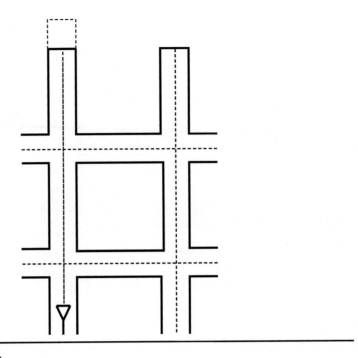

Illustrates Problem 9.1

Loading rate = 10 tons per min
Coal density = 0.04 tons per cu ft
Shuttle-car discharge time = 0.5 min
Shuttle cars = 25-ft long
Belt feeder (dump) located 110 ft outby the change point
Switch-in time = 0.15 min
Switch-out time = 0.15 min

Problem 9.2

Assume that shuttle car B was in need of repair at the start of the cycle of the previous problem and the decision was made to load out the cut with only shuttle car A. Determine how long it would take to load out the cut.

Problem 9.3

For the conventional-mining plan shown, determine the time required to load out cut no. 80 (i.e., put all of the coal on the panel belt). For tram-distance calculations, measure from the midpoint of the cut. List the operating times for each shuttle car (one column for car B and one column for car C). Neglect the time it takes to tram the loader to cut no. 81. Pertinent information:

Shuttle-car tram rate (empty) = 250 fpm
Shuttle-car tram rate (loaded) = 225 fpm
Shuttle-car payload = 6 tons
Loading machine = 20-ft long

RUBBER-TIRED HAULAGE SYSTEMS | 253

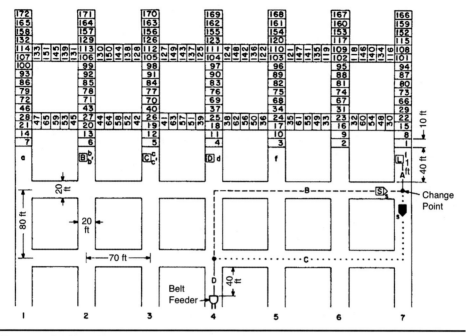

Illustrates Problem 9.3

Loading rate = 4 tons per min
Coal density = 0.04 tons per cu ft
Seam thickness = 6 ft
Shuttle-car discharge time = 0.5 min
Shuttle cars = 25-ft long

Problem 9.4

Pictorially describe the answer to Problem 9.3 with bar graphs representing the cycle times for cars B and C.

Problem 9.5

Verify the answers to Problems 9.3 and 9.4 with the equations provided for the mathematical modeling of face-haulage systems.

Problem 9.6

Two battery-powered haulage vehicles are projected to replace the shuttle cars in the section layout shown in problem 9.3. Within which cut in entry 7 will WSC time be encountered because no empty vehicle is available at the cleared change point? Assume the following:

Change point remains as shown.
Loader length is 20 ft.
Neglect the length of the battery-operated haulage vehicles.

Turn time at the change point and the dump is 0.25 min per occurrence.
Loading time is 1.75 min.
Tram rate (loaded) is 250 fpm.
Tram rate (empty) is 325 fpm.
Discharge time is 0.25 min.

Problem 9.7

Utilizing the same change point as shown in the layout for problem 9.3, what would be the minimum loading rate of the loader to maintain a shuttle-car zero wait time (i.e., no WSC time) for cut no. 80 if the shuttle-car speeds are 325 fpm and each car has a 10-ton capacity? Assume that the mean dump time is 0.5 min. and the loader is 20 ft long.

Problem 9.8

A boring-type continuous miner is to be used to advance an entry in a 6-ft seam of potash under a self-supporting roof. The distance from the change point to the dump is 280 ft for both of the 5-ton capacity shuttle cars, which travel at an average speed of 300 fpm. The loading rate is 3.2 tons per min, and the shuttle car dump time is 30 sec. If each foot of miner advance produces 4 tons of potash, determine how far the continuous miner can advance the entry beyond the change point before the additional wait time (WSC) is eliminated. Assume that the breakthroughs and entries are driven 18-ft wide and that neither shuttle car can enter the intersection containing the change point until it is cleared by the other car (as shown in Figure 9.2). The lengths of the continuous miner and shuttle cars are 23 ft and 25 ft, respectively.

Problem 9.9

(Figures 9.3 to 9.10 and Tables 9.1 and 9.2 should be used for Problems 9.9 through 9.11.)

Estimate the cycle time and production of a 150,000-lb GVW off-highway truck with 80,000 lb on its rear wheels when loaded to its rated capacity, if it operates on a 5,000-ft level haul road. The road flexes under load, is rutted, and has little maintenance. Assume the following:

Loading time: 1.30 min
Maneuver and dumping time: 0.75 min
Job efficiency: 50 min. of operating time per scheduled hour (0.83)

Problem 9.10

The surface-mine engineer for the operation described in problem 9.9 is considering the use of an alternate haul road for the off-highway trucks. Although the alternative is shorter (3,000 ft), it has a 5% adverse grade on the haul. Assume that all other concerns (loading time, maneuver and dumping time, job efficiency, coefficient of traction, rolling resistance, etc.) are identical to those of problem 9.9. Which haul route (the one in the previous problem or the one in this problem) should be selected based on lowest cycle times? What are the cycle times and hourly productions?

Problem 9.11

If the truck described in problem 9.10 encounters a 1500-ft 20% downgrade while on the haul, what is its maximum safe sustainable speed so that its brakes can function without exceeding their cooling capacity? How long will it take for the vehicle to travel this distance?

Problem 9.12

The 150,000-lb GVW off-highway truck described in the previous problems is rated at 450 hp (flywheel power). What percent of this flywheel power is available at an elevation of 7,500 ft?

PROBLEM SOLUTIONS

Solution 9.1

Tons in cut no. 80: 6 × 20 × 20 × 0.04 = 96 tons
Number of shuttle-car loads: 96 ÷ 8 = 12
Loading time: 8 ÷ 10 = 0.8 min per car
Change-out distance (COD) = [(0.5)(entry width)] + (change point to midpoint of cut) − (miner length) = (10 ft) + (80 ft) − (25 ft) = 65 ft
Haul distance (HD) to the dump = [(distance from the point where the "imaginary dot" on the inby end of the shuttle car clears the intersection containing the change point to the dump) − (shuttle car length) = 100 ft − 25 ft = 75 ft
Dump time: 0.5 min
Travel time (loaded and empty): COD = 65/300 = 0.22 min
HD = 75/300 = 0.25 min
The inby cycle times are greater than the outby cycle times:

$$SW\text{-}OUT + COD + LOAD + COD > HD + DUMP + HD + SW\text{-}IN$$

$$0.15 + 0.22 + 0.80 + 0.22 > 0.25 + 0.50 + 0.25 + 0.15$$

$$1.39 > 1.15$$

Therefore, there will always be an empty shuttle car waiting for 0.24 min at the change point and WSC will equal zero for the entire cycle.

Cycle Times

	Car A			Car B
0.80	Load No. 1		0.80	WAIT
0.22	COD		0.22	WAIT
			0.15	SWITCH-OUT
0.25	HD		0.22	COD
0.50	Dump No. 1		0.80	Load No. 2
0.25	HD		0.22	COD
0.15	SWITCH-IN		0.25	HD
0.24	WAIT		0.50	Dump No. 2
0.15	SWITCH-OUT		0.25	HD

(table continues on next page)

Cycle Times (continued)

Car A		Car B	
0.22	COD	0.15	SWITCH-IN
0.80	Load No. 3	0.24	WAIT
0.22	COD	0.15	SWITCH-OUT
0.25	HD	0.22	COD
0.50	Dump No. 3	0.80	Load No. 4
0.25	HD	0.22	COD
0.15	SWITCH-IN	0.25	HD
0.24	WAIT	0.50	Dump No. 4
0.15	SWITCH-OUT	0.25	HD
0.22	COD	0.15	SWITCH-IN
0.80	Load No. 5	0.24	WAIT
0.22	COD	0.15	SWITCH-OUT
0.25	HD	0.22	COD
0.50	Dump No. 5	0.80	Load No. 6
0.25	HD	0.22	COD
0.15	SWITCH-IN	0.25	HD
0.24	WAIT	0.50	Dump No. 6
0.15	SWITCH-OUT	0.25	HD
0.22	COD	0.15	SWITCH-IN
0.80	Load No. 7	0.24	WAIT
0.22	COD	0.15	SWITCH-OUT
0.25	HD	0.22	COD
0.50	Dump No. 7	0.80	Load No. 8
0.25	HD	0.22	COD
0.15	SWITCH-IN	0.25	HD
0.24	WAIT	0.50	Dump No. 8
0.15	SWITCH-OUT	0.25	HD
0.22	COD	0.15	SWITCH-IN
0.80	Load No. 9	0.24	WAIT
0.22	COD	0.15	SWITCH-OUT
0.25	HD	0.22	COD
0.50	Dump No. 9	0.80	Load No. 10
0.25	HD	0.22	COD
0.15	SWITCH-IN	0.25	HD
0.24	WAIT	0.50	Dump No. 10
0.15	SWITCH-OUT	0.25	HD
0.22	COD	0.15	SWITCH-IN
0.80	Load No. 11	0.24	WAIT
0.22	COD	0.15	SWITCH-OUT
0.25	HD	0.22	COD
0.50	Dump No. 11	0.80	Load No. 12
		0.22	COD
		0.25	HD
		0.50	Dump No. 12 (contains the last ton of coal)

Cycle Time for Car B:

$$0.80 + 0.22 + 0.15 + 0.22 + 0.80 + 0.22 + 0.25 + 0.50 + [5\,(0.25 + 0.15 + 0.24 + 0.15$$
$$+ 0.22 + 0.80 + 0.22 + 0.25 + 0.50)] = 17.06 \text{ min}$$

Check with the Mathematical Model:

The mathematical model to be used, with the continuous miner tram time ignored, is:

$$LCT = LT + COT + WSC$$

Further:

$$LT = (T/LR) = (96/10) = 9.6 \text{ min}$$

Because $N = (T/CAP)^* = (96/8)^* = (12)^* = 13$, COT becomes:

$$COT = (13 - 1)(2\,COD/SPD) = (13-1)(2\,(65/300)) = 5.2 \text{ min}$$

This equation for COT assumes that the first haulage vehicle must tram inby the change point, and this time (0.22 min) is included in the 5.2 min. Since this problem assumes that the cycle begins with the haulage vehicle already at the miner, the true value of COT is 4.98.

Finally, WSC has already been shown to equal zero. However, because of the "switch-out" cycle, the model has to accommodate a time of 0.15 min per occurrence (12 in all). We can include these 1.80 min and treat them as WSC time because the miner is ready, but a shuttle car is not available until the "switch-out" operation takes place. Therefore,

$$LCT = 9.6 + 4.98 + 1.80 = 16.38 \text{ min}$$

Since the mathematical model assumes that the cycle ends when the last loaded haulage vehicle clears the change point, the time for haulage from the change point to the dump (0.25) and the time to dump the last load (0.5) must be added to 14.80 min. The net result is $16.38 + 0.25 + 0.5 = 17.13$ min, which, with round-off error, is close to the answer determined earlier.

Solution 9.2

Unconstrained by the presence of another shuttle car, the cycle for car A begins as:

Cycle Times

\	Car A
0.80	Load No. 1
0.22	COD
0.25	HD
0.50	Dump No. 1
0.25	HD
0.22	COD
0.80	Load No. 2
0.22	COD
0.25	HD
0.50	Dump No. 2

It should be readily apparent that one cycle (from the completion of dump no. 1 to the completion of dump no. 2) takes 2.24 min (0.25 + 0.22 + 0.80 + 0.22 + 0.25 + 0.50). Therefore, we can revise the cycle times to the following format:

Cycle Times

	Car A
0.80	Load No. 1
0.22	COD
0.25	HD
0.50	Dump No. 1
2.24	Complete Cycle to Dump No. 2
2.24	Complete Cycle to Dump No. 3
2.24	Complete Cycle to Dump No. 4
2.24	Complete Cycle to Dump No. 5
2.24	Complete Cycle to Dump No. 6
2.24	Complete Cycle to Dump No. 7
2.24	Complete Cycle to Dump No. 8
2.24	Complete Cycle to Dump No. 9
2.24	Complete Cycle to Dump No. 10
2.24	Complete Cycle to Dump No. 11
2.24	Complete Cycle to Dump No. 12

Cycle Time for Car A:

$$0.80 + 0.22 + 0.25 + 0.50 + [11 (2.24)] = 26.41 \text{ min}$$

In other words, one car working alone can load out this cut at approximately 65% of the two-car rate.

Solution 9.3

Tons in cut no. 80: $6 \times 20 \times 10 \times 0.04 = 48$
Number of cars: $48 \div 6 = 8$
Loading time: $6 \div 4 = 1.5$ min per car
COD = [(0.5)(entry width)] + (change point to midpoint of cut) − (loader length) =
  (10 ft) + (10 ft + 40 ft + 70 ft + 5 ft) − (20 ft) = 115 ft
HD to the dump = 70 ft + 70 ft + 70 ft + 70 ft + 50 ft − 25 ft = 305 ft
Dump time: 0.5 min
Travel time (loaded): COD = 115/225 = 0.51 min
HD = 305/225 = 1.36 min
Travel time (empty): HD = 305/250 = 1.22 min
COD = 115/250 = 0.46 min

The outby cycle times are greater than the inby cycle times:

$$COD + LOAD + COD < HD + DUMP + HD$$

$$0.46 + 1.50 + 0.51 < 1.36 + 0.50 + 1.22$$

$$2.47 < 3.08$$

Therefore, a WSC time will be observed.

Cycle Times

	Car B		Car C
1.50	Load No. 1	1.50	WAIT
0.51	COD	0.51	WAIT
1.36	HD	0.46	COD
0.50	Dump No. 1	1.50	Load No. 2
1.22	HD	0.51	COD
0.46	COD	1.36	HD
1.50	Load No. 3	0.50	Dump No. 2
0.51	COD	1.22	HD
1.36	HD	0.46	COD
0.50	Dump No. 3	1.50	Load No. 4
1.22	HD	0.51	COD
0.46	COD	1.36	HD
1.50	Load No. 5	0.50	Dump No. 4
0.51	COD	1.22	HD
1.36	HD	0.46	COD
0.50	Dump No. 5	1.50	Load No. 6
1.22	HD	0.51	COD
0.46	COD	1.36	HD
1.50	Load No. 7	0.50	Dump No. 6
0.51	COD	1.22	HD
1.36	HD	0.46	COD
0.50	Dump No. 7	1.50	Load No. 8
		0.51	COD
		1.36	HD
		0.50	Dump No. 8 (contains the last ton of coal)

Cycle Times for Car C:

$$1.5 + 0.51 + 3(5.55) + 0.46 + 1.5 + 0.51 + 1.36 + 0.50 = 22.99 \text{ min}$$

Solution 9.4

The following shadings will represent the various cycle times for the haulage vehicles:

Loading Time (1.5 min)
Change-Out Time/Loaded (0.51 min)
Haul Time/Loaded (1.36 min)
Dump (0.50 min)
Haul Time/Empty (1.22 min)
Change-Out Time/Empty (0.46 min)
Wait at Change Point (varies)

Illustrates Solution to Problem 9.4

260 | MINING ENGINEERING ANALYSIS

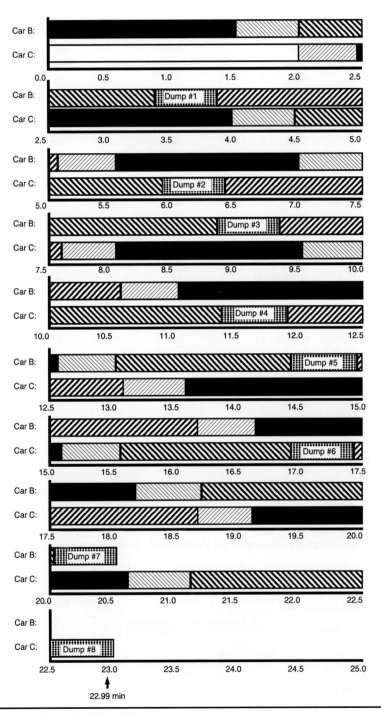

Illustrates Solution to Problem 9.4

Solution 9.5

The mathematical model to be used, with loader tram time ignored, is:

$$LCT = LT + COT + WSC$$

Further:

$$LT = \frac{T}{LR} = \frac{48}{4} = 12 \text{ min}$$

Because this problem has two shuttle-car speeds, an empty speed of 250 fpm, and a loaded speed of 225 fpm, it is appropriate to substitute the average speed, 237.5 fpm, for SPD in the model. Further, because $N = (T/CAP)^* = (48/6)^* = (8)^* = 9$, COT becomes:

$$COT = (N-1)\left(\frac{2COD}{SPD}\right) = (9-1)2\left(\frac{115}{237.5}\right) = 7.75 \text{ min}$$

This equation for COT assumes that the first haulage vehicle must tram inby the change point, and this time (0.46 min) is included in the 7.75 min. Since this problem assumes that the cycle begins with the haulage vehicle already at the loader, the true value of COT is 7.29.

Finally, WSC is calculated as follows:

$$WSC = \left[\frac{N-1}{2}\right]_* \left[\left(\frac{2HD}{SPD}\right) + DT\right] - \left[(NO-1)\left(\frac{CAP}{LR}\right) + \frac{2COD}{SPD}\right]$$

$$= \left[\frac{9-1}{2}\right]_* [2.568 + 0.5] - \left[(2-1)\left(\frac{6}{4}\right) + (0.968)\right]$$

$$= (3)(2.568 + 0.5) - (2.468)$$

$$= (3)(0.6)$$

$$= 1.80 \text{ min}$$

Therefore,

$$LCT = 12 + 7.29 + 1.80 = 21.09 \text{ min}$$

Because the mathematical model assumes that the cycle ends when the last loaded haulage vehicle clears the change point, the time for haulage from the change point to the dump (1.36) and the time to dump the last load (0.5) must be added to 21.09 min. The net result is 21.09 + 1.36 + 0.5 = 22.95 min, which, with round-off error, is close to the answer determined in the previous two problems.

Solution 9.6

Cleared change point to dump: 70 ft + 70 ft + 70 ft + 70 ft + 40 ft = 320 ft
Outby cycle time: (320/250) + 0.25 + 0.25 + 0.25 + (320/325) = 3.01 min
Inby cycle time: cleared change point to loader (empty) + loading + loader to cleared change point (loaded) = [X ft (0.003 min per ft)] + (1.75 min) + [X ft (0.004 min per ft)]

Therefore, the break-even point is:

$$3.01 \text{ min} = [X \text{ ft } (0.003 \text{ min per ft})] + (1.75 \text{ min}) + [X \text{ ft } (0.004 \text{ min per ft})]$$

$$1.26 \text{ min} = X(0.004 + 0.003)$$

$$X = 180 \text{ ft}$$

A tram distance of 180 ft inby the cleared change point, plus an additional 20 ft due to the length of the loader, represents the start of cut no. 159. Thus, WSC time should be eliminated during the mining of cut no. 159.

Solution 9.7

The easiest way to solve this problem is to equate the inby-the-cleared-change-point and outby-the-cleared-change-point haulage times:

$$COT = 2 \text{ (COD/SPD)} = 2 [(140 - 20)/325] = 0.7385 \text{ min}$$

COT + LT = tram time to the dump + dump time + travel time from the dump

$$0.7385 + LT = \left(\frac{320}{325}\right) + 0.5 + \left(\frac{320}{325}\right)$$

$$LT = 1.73 \text{ min}$$

$$LT = \frac{CAP}{LR}$$

Thus,

$$LR = \frac{CAP}{LT} = \frac{10}{1.73} = 5.78 \text{ tons per min}$$

Solution 9.8

Outby tram time:

$$280 \text{ ft} - \left(\frac{18}{2}\right) \text{ ft} - 25 \text{ ft} = 246 \text{ ft}$$

$$(2)\left(\frac{246}{300}\right) + 0.5 = 2.14 \text{ min}$$

Inby tram time:

$$(2)\frac{(9 + 9 + (X - 23) \text{ ft})}{300 \text{ fpm}} + \left(\frac{5 \text{ tons}}{3.2 \text{ tons per min}}\right) = 2.14 \text{ min}$$

$$(2)\left(\frac{-5 + X}{300}\right) \text{min} + 1.56 \text{ min} = 2.14 \text{ min}$$

$$X = 92 \text{ ft from inby rib to face}$$

$$\text{CP to face} = 92 + 9 = 101 \text{ ft}$$

Solution 9.9

At full load, truck haul: 81,000 lb (150,000 lb − 68,900 lb) or 40.55 tons
Grade: 0% because it is a level haul road
Rolling resistance: 100/20 = 5% (Table 9.1)
Total effective grade: 5% + 0% = 5%
Cycle time (5,000 ft = 1524 m):

Load time:	1.30 min
Dump, maneuver time:	0.75 min
Haul time:	2.80 min (Figure 9.4)
Return time:	1.60 min (Figure 9.5)
	6.45 min per cycle

$$\text{Cycles per hour} = [60 \text{ min per hr } (0.83)]/6.45 \text{ min per cycle} = 7.72$$

$$\text{Tonnage hauled per hour} = (7.72)(40.55) = 313 \text{ tons}$$

264 | MINING ENGINEERING ANALYSIS

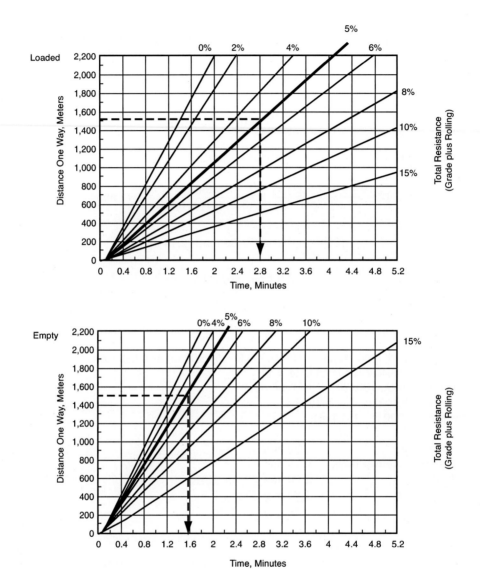

Illustrates Solution to Problem 9.9

Solution 9.10

Problem 9.9 established the cycle time of 7.72 min. What was not determined was whether or not the vehicle could operate under the given conditions:

Rolling resistance factor: 5%
Grade resistance on haul: 5%
Grade resistance on return: −5%
Effective grade on haul: 10%
Effective grade on return: 0%
Gross vehicle weight on haul: 150,000 lb
Gross vehicle weight on return: 150,000 − 81,100 = 68,900 lb
Power required on the haul: 150,000 × 0.10 = 15,000 lb
Power required on the return: 68,900 × (0.00) = 0.0 lb
The available power can be determined using Figure 9.3: 15,000 lb @ 8.0 mph.

Thus, the vehicle should be able to operate under the given conditions.
Cycle time (3,000 ft = 914.4 m):

Load time:	1.30 min
Dump, maneuver time:	0.75 min
Haul time:	3.35 min (Figure 9.4)
Return time:	0.75 min (Figure 9.5)
	6.15 min per cycle

Cycles per hour = [60 min per hr (0.83)]/[6.15 min per cycle] = 8.10

Tonnage hauled per hour = (8.10)(40.55) = 329 tons

In summary, the new situation should be selected, because the cycle time is faster (6.15 min versus 6.45 min) and the hourly production is higher (329 tons versus 313 tons) than the situation described in problem 9.9.

266 | MINING ENGINEERING ANALYSIS

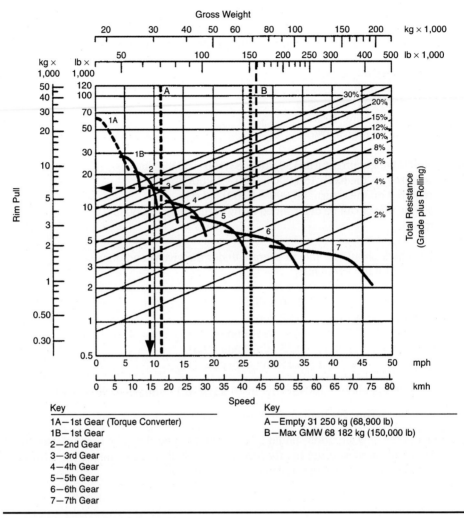

Key
1A—1st Gear (Torque Converter)
1B—1st Gear
2—2nd Gear
3—3rd Gear
4—4th Gear
5—5th Gear
6—6th Gear
7—7th Gear

Key
A—Empty 31 250 kg (68,900 lb)
B—Max GMW 68 182 kg (150,000 lb)

Illustrates Solution to Problem 9.10

SURFACE EXTRACTION

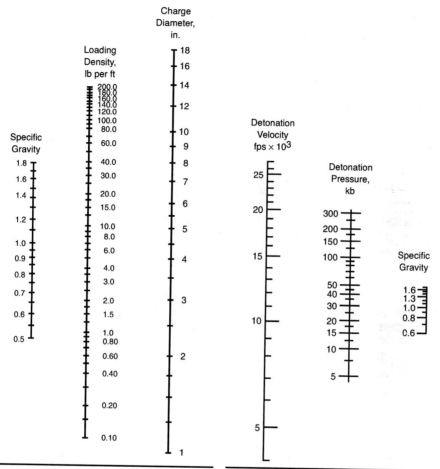

FIGURE 10.1 Nomograph for determining loading density
Source: Dick 1973.

FIGURE 10.2 Nomograph for determining detonation pressure
Source: Dick 1973.

blasting agents are the most common in coal mining because they are safe to use, due to the fact that they need a primer for detonation and are relatively inexpensive. Ammonium nitrate/fuel oil has become the most common dry blasting agent.

DESIGN OF BLASTING ROUNDS

The proper design of a blasting round for a surface mine or quarry requires a balance among overburden characteristics, regulatory requirements, explosive technology, and overall economics. Figure 10.3 shows the various parameters of a blasting round.

Two common approaches are used to design blasting rounds. One approach is driven by legal requirements and concern for ground vibration, while the other reflects empirical data taken from successful midwestern quarries.

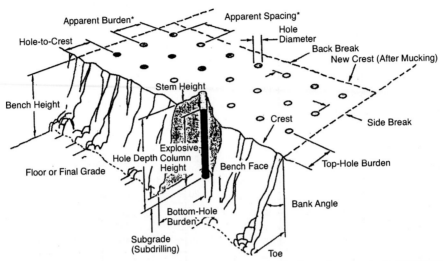

FIGURE 10.3 Parameters of a blasting round
Source: Dyno Nobel 1997.

The equation that drives the ground-vibration approach is:

$$D = (Ds)(\sqrt{W}) \qquad \text{(EQ 10.2)}$$

where: D is the distance (in feet) to the structure to be protected, Ds is the scaled distance, and W is the weight of explosives allowed per 8 m-sec (millisecond) delay (Austin Powder 2002). Figure 10.4 shows the federal standards (30 CFR 816.67(d)(4)) relating maximum allowable particle velocity (V) to blast vibration frequency.

Figures 10.5 and 10.6 show some of the characteristics and parameters used in the design of blasting rounds by the empirical method.

Figure 10.5 shows those dimensions that are considered for a single hole, while Figure 10.6 gives dimensions for a total round, including:

D_e = diameter of the explosive in the borehole

B = burden, or the distance between the rows of holes running parallel to the free vertical surface of the rock

S = spacing, or distance between two holes in each row

H = length of the hole

J = subdrilling length, or the length the hole is drilled below the floor of the working pit

T = the portion of the hole that does not contain explosive, or collar length

L = bench height

b = spacing between rows of holes as measured perpendicular to the free face

s = separation between adjacent holes in a given row

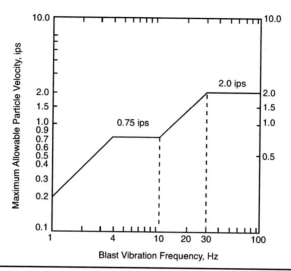

FIGURE 10.4 Federal standards for blast vibration
Source: Office of Surface Mining 2001.

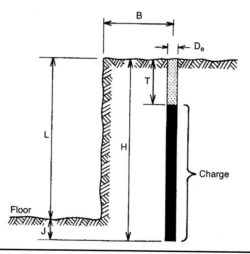

FIGURE 10.5 Bench cross section view showing D_e, B, H, J, T, and L
Source: Ash 1990.

It should be pointed out that b and s are terms used by operators when referring to the drilling grid; they should not be confused with B and S, the burden and spacing that are measured with reference to free faces.

The following five ratios should be used in the design of rounds. Assuming that all measurements of length and diameter are expressed in the same units (inches, feet, etc.), the ratios relate the dimensions to be designed through the following set of equations:

276 | MINING ENGINEERING ANALYSIS

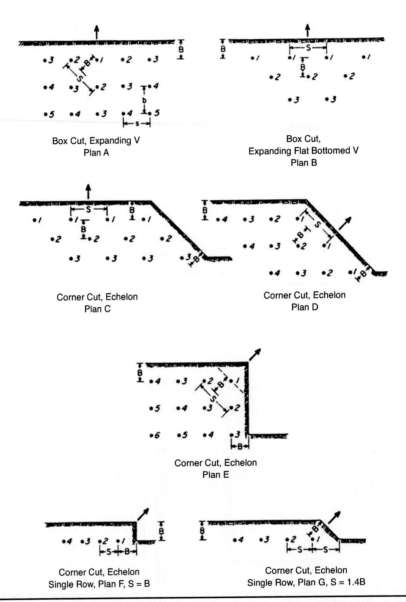

FIGURE 10.6 Generalized blasting patterns showing B, S, b, and s (numbers indicate firing sequence)
Source: Pugliese 1972.

$$B = K_B D_e; \quad K_B = \text{burden ratio} \quad \text{(EQ 10.3)}$$

$$S = K_S B; \quad K_S = \text{spacing ratio} \quad \text{(EQ 10.4)}$$

$$H = K_H B; \quad K_H = \text{hole length ratio} \quad \text{(EQ 10.5)}$$

$$J = K_J B; \quad K_J = \text{subdrilling ratio} \quad \text{(EQ 10.6)}$$

$$T = K_T B; \quad K_T = \text{collar distance ratio} \quad \text{(EQ 10.7)}$$

From these relationships, all the critical dimensions of blasting-round design can be calculated if appropriate values for the various K ratios are selected. Although these relationships were developed in studies of limestone quarrying, the relationships still hold for surface coal and other open-pit operations, and the K values that have been developed are valid for first assumptions. As is the case with any empirically based design criteria, empirical relationships from the literature are usually accepted as only a first approximation. Subsequent iterations resulting in more exact relationships can be developed as actual practice of these design steps is followed.

For calculating the burden, given a particular D_e, the following ratios are considered to be appropriate for limestone and dolomite:

K_B = 30, under average conditions
K_B = 25, for low-density explosives such as ANFO
K_B = 35, for high-density slurry explosives and gelatins

The actual K_B ratios are also related to the density of the rock. Limestone and dolomite have a density of approximately 2.7 g/cm^3. Different values of K_B should be used for rocks with densities greater than 2.7 (e.g., 2.85 and above) or for low densities (e.g., 2.55 and below).

The calculation of spacing is based on K_S values between 1.8 and 2.0 for simultaneously initiated holes in a given row. The most favorable conditions are found with staggered drill patterns and rows in which all charges are not detonated simultaneously. If exact timing is achieved, simultaneously fired holes can be designed using K_S values of 3 to 5. These charges must be sufficiently long and exactly timed to allow correct enhancement of stress effects.

Spacing is also determined with respect to height and burden relationships, specifically the ratio H/B. The following relationships are recommended:

$$S = (BH)^{0.5} \quad \text{for } 2B \leq H < 4B \tag{EQ 10.8}$$

$$S = 2B \quad \text{for } 4B \leq H \tag{EQ 10.9}$$

In this situation, the K_S of 1.8 to 2.0 is satisfactory, although lower K values can be assigned with H/B less than 3.0.

If delays are sequenced in a given row, the K_S should range between 1.0 and 1.2. Square drilling patterns are preferred and will cause the rock to expand in a wedge or pattern aligned at 45° from the original open face. Considering the ratio of s and b, where b = 1.4B for the regular case, the design relationship,

$$S = K_S(b) = K_S(1.4B) \tag{EQ 10.10}$$

is appropriate where K_S = 1.0 to 1.2.

The values of the ratio K_S may be adjusted between 1 and 2 to meet local conditions, particularly the period of delay between charges. Large ratios create spacing so much greater than the burden that the vertical face craters and lumps are left on the pit floor. Small values of K_S, less than 1, cause excessive breakage, premature breakage between holes, and progressive loss from the hole, creating slabs, boulders, and toe problems. The nature of these particular relationships is such that they assume ideal energy balance between charges, so further experimentation with the balance of individual holes may be indicated if such problems are not resolved by varying the K_S ratio.

TABLE 10.1 Worksheet for design of blasting rounds

$D_e =$ _____ in.; Cut _____; Desired rock progression

$B = K_B D_e/12 =$ _____ ft for $K_B \approx 30$ (average)

$B = K_B D_e/12 =$ _____ ft for $K_B \approx$ ___ (alternative)

For staggered pattern, simultaneous timing

$S = K_B B$ = _____ ft for $K_S \approx 2$ (average)

$S = K_B B$ = _____ ft for $K_S \approx 1.8$ (alternative)

For square pattern, sequence timing

$S = K_S b = K_S (1.4\,B)$ = _____ ft for $K_S \approx 1$ (average)

$S = K_S b = K_S (1.4\,B)$ = _____ ft for $K_S \approx 1.2$ (alternative)

$H = K_H B$ = _____ ft for $K_H \approx 2.6$ (average)

$H = K_H B$ = _____ ft for $K_H \approx$ ___ (alternative)

$H_{min} = K_H B$ = _____ ft for $K_H \approx 1.5$ (minimum)

$H_{max} = K_H B$ = _____ ft for $K_H \approx 4$ (maximum)

$J = K_J B$ = _____ ft for $K_J \approx 0.3$ (average)

$J = K_J B$ = _____ ft for $K_J \approx$ ___ (alternative)

$T = K_T B$ = _____ ft for $K_T \approx 0.7$ (average)

$T = K_T B$ = _____ ft for $K_T \approx$ ___ (alternative)

$L \approx H - J$ = _____ ft (average or alternative)

$L_{min} \approx H_{min} - J$ = _____ ft (minimum)

$L_{max} \approx H_{max} - J$ = _____ ft (maximum)

Source: Pugliese 1972.

Hole length calculations are generally based on K_H ratios of 1.5 to 4.0, with 2.6 representing the typical K_H value. The depth should almost always be greater than or equal to the burden to reduce overbreak and cratering tendencies. Conversely, high values, greater than 4.0, may create underbreak or bootlegging if the hole is single primed. Staged priming may allow lengths greater than those indicated by K_H values of 4.0. No definite limit to K_H can be established without a definite examination of the cratering properties of the specific material being blasted.

The subdrilling length, which is applicable only to quarries and not to coal mines, is generally indicated by K_J values greater than 0.2; a value of 0.3 is suggested as preferable to ensure a full face and quarry floor. In the case of overburden drilling, however, care should be taken to limit fracturing of the coal seam and subsequent unnecessary removal of coal with the overburden. In some instances, negative K_J values may be applied. Common practice is to drill to the coal seam and then backfill by 3 to 5 ft.

The collar distance can be calculated by using a K_T value of 0.7. The uncharged portion of the hole is usually filled with stemming material and is left uncharged to prevent air blast and promote gas confinement within the borehole. In solid rock, it is suggested that values of K_T less than 1.0 can lead to cratering, backbreak, and violence, particularly with collar priming.

The various calculations are presented in work sheet form in Table 10.1. Once the original face and the shape of the cut are selected, the required geometrical relationships can be determined from the plan layouts of the cuts. For any given explosive type, blast hole diameter, and rock density, the burden is calculated from Eq. 10.3. Equation 10.4 is then solved for the spacing. The hole length can be checked according to Eq. 10.5 and the timing, delay, and sequencing constraints that the operator has established. Finally, the subdrilling length (Eq. 10.6), if necessary, can be calculated, and the collar distance can be determined (Eq. 10.7).

NUMBER OF DRILLS

In surface mining, drilling and blasting is performed to remove overburden. The amount of drilling and blasting depends on the types of strata present in the area; hard rocks require more drilling and blasting. The number of drills needed for drilling can be calculated as follows:

SYE = square yards of bench to be exposed

COB = cubic yards of overburden removed per month

DS = spacing of the drill holes in feet

DB = burden of the drill holes in feet

DP = drilling pattern in square feet

LDPM = length of drilling, in feet, required per month

NH = number of drill holes

DR = drilling rate in feet per hour

HPS = hours scheduled per shift

SPD = shifts per day

DPW = drilling days per week

SPM = weeks worked per month

SMH = scheduled monthly hours

DPM = number of drills required

$$DP = (DS)(DB) \qquad \text{(EQ 10.11)}$$

$$NH = \frac{(SYE)(9)}{DP} \qquad \text{(EQ 10.12)}$$

$$LDPM = \left(\frac{COB}{DP}\right)(27) \qquad \text{(EQ 10.13)}$$

$$SMH = (HPS)(SPD)(DPW)(SPM) \qquad \text{(EQ 10.14)}$$

$$DPM = \frac{LDPM}{(SMH)(DR)} \qquad \text{(EQ 10.15)}$$

AMOUNT OF EXPLOSIVES REQUIRED

Fundamental to effective blasting is the use of the correct amount of explosive for the volume of overburden to be fragmented. This relationship is usually expressed as the powder factor (PF), defined as the pounds of explosives used per hole divided by the cubic yards of overburden fragmented per hole (Austin Powder 2002). A typical value would be 1.1 for strong sandstones.

To determine the PF, use the following equation:

$$PF = \frac{(27)(W)}{(B)(S)(H)} \qquad \text{(EQ 10.16)}$$

where:

    W = pounds of explosive per hole

    B = burden (in feet)

    S = spacing (in feet)

    H = bench height (in feet)

Therefore, knowing the number of holes drilled per month, the drilling pattern, and the overburden depth, the amount of explosive required per month can be easily calculated.

Oftentimes, however, not all the overburden requires fragmentation before removal. This is particularly true for topsoil and soft soil. Therefore, actual consumption of explosives must be calculated on the basis of experience with the overburden; in other words, that fraction of the overburden that should be fragmented must be known.

EXCAVATION FUNDAMENTALS

The first step in excavation fundamentals is to determine bucket capacity. Calculations are conducted in bank volumes, in other words, the material to be excavated is in the solid or in situ state. Bucket capacity is determined as follows:

$$Bc = \frac{Q}{(C)(S)(A)(O)(Bf)(P)} \qquad \text{(EQ 10.17)}$$

Since Eq. 10.17 contains a significant number of parameters that must be defined, refer to Table 10.2 for a thorough description of each.

When using manufacturers' or other tabled factors, be careful that you know the full meaning of each parameter. For example, propel time, P, is occasionally combined with O or OA; to avoid double compensation in those instances, be sure to remove P from Eq. 10.17. Further, in those instances where O and A cannot be determined, their product (OA: operating efficiency) can be determined from Table 10.3.

DRAGLINE SELECTION

Draglines work on a bench on the highwall or on top of the highwall itself. An imaginary vertical plane passing through the edge of the highwall would contain the point D, shown to the left of the tub in Figure 10.7. The projection of a horizontal line from the

TABLE 10.2 Bucket-capacity parameters

Parameter	Description
Bc	Bucket capacity (bank volume)
Q	Production required (bank volume per hour)
C	Theoretical cycles per hour for a 90° swing; C = 60/tc, where tc is the time, in minutes, it takes the shovel to swing 90°. Values of C may also be obtained from manufacturers' literature or from time studies.
A	Mechanical availability during the scheduled hours of work.
O	Job operational factor. Since the shovel is part of a system, corrections must be made when there are delays due to management and labor deficiencies, job conditions, climate, etc.
Bf	Bucket factor (Bf). Bf can also be described in the following manner: Bf = (fillability)/(swell factor). Fillability is the loose volume of material excavated in an average load as a ratio of the bucket capacity, Bc; although fillability can be determined from field measurements, Table 10.4 can also provide an approximate guide. Swell factor is defined as follows: Swell factor = [weight per unit volume (bank)]/[weight per unit volume (loose)]. The reciprocal of swell factor is the percent swell.
S	Swing factor. Shovel cycle times are usually based on a 90° swing. Table 10.5 provides corrections for other angles.
P	Propel time factor. This factor takes into account the time it takes a shovel to maneuver and can usually be determined through time studies. Table 10.6 provides typical values for this factor.

Source: Atkinson 1992.

TABLE 10.3 Shovel operating efficiency

Job Conditions	Management Conditions			
	Excellent	Good	Fair	Poor
Excellent	0.83	0.80	0.77	0.70
Good	0.76	0.73	0.70	0.64
Fair	0.72	0.69	0.66	0.60
Poor	0.63	0.61	0.59	0.54

Source: Atkinson 1992.

imaginary vertical plane to the point of discharge of the bucket contents establishes the reach of the dragline. Using Figure 10.7, it can be shown that the reach of a dragline is:

$$Rd = \frac{H}{\tan\phi} + \frac{1}{\tan\theta}\left[H\left(1 + \frac{SP}{100}\right) + \left(\frac{W}{4}\tan\theta\right) - T\right] \quad \text{(EQ 10.18)}$$

where Rd is the reach of the dragline in feet, and SP is the swell in percent. If E_t is defined as the outside diameter of the dragline tub, and it is assumed that the dragline is offset from the highwall for a distance of three quarters of the tub diameter, the dumping radius (Rdd) is as follows:

TABLE 10.4 Bank density, swell factor, and fillability of common materials

Rock	Bank Density (lb per cu yd)	Swell Factor	Fillability
Asbestos ore	3,200	1.4	0.85
Basalt	5,000	1.6	0.80
Bauxite	3,200	1.35	0.90
Chalk	3,100	1.3	0.90
Clay (dry)	2,400	1.25	0.85
Clay (light)	2,800	1.3	0.85
Clay (heavy)	3,600	1.35	0.80
Clay and gravel (dry)	2,500	1.3	0.85
Clay and gravel (wet)	3,000	1.35	0.80
Coal (anthracite)	2,700	1.35	0.90
Coal (bituminous)	2,100	1.35	0.90
Coal (lignite)	1,700	1.3	0.90
Copper ores (low-grade)	4,300	1.5	0.85
Copper ores (high-grade)	5,400	1.6	0.80
Earth (dry)	2,800	1.3	0.95
Earth (wet)	3,400	1.3	0.90
Granite	4,000	1.55	0.80
Gravel (dry)	3,000	1.25	1.0
Gravel (wet)	3,600	1.25	1.0
Gypsum	4,700	1.5	0.85
Limonite	5,400	1.4	0.85
Iron ore (40% Fe)	4,500	1.4	0.80
Iron ore (+40% Fe)	5,000	1.45	0.80
Iron ore (+60% Fe)	6,500	1.55	0.75
Iron ore (taconite)	8,000	1.65	0.75
Limestone (hard)	4,400	1.6	0.80
Limestone (soft)	3,700	1.5	0.85
Manganese ore	5,200	1.45	0.85
Phosphate rock	3,400	1.5	0.85
Sand (dry)	2,900	1.15	1.00
Sand (wet)	3,400	1.15	1.00
Sand and gravel (dry)	3,300	1.15	1.00
Sand and gravel (wet)	3,800	1.15	1.00
Sandstone (porous)	4,200	1.6	0.80
Sandstone (cemented)	4,500	1.6	0.80
Shale	4,000	1.45	0.80

Source: Atkinson 1992.

TABLE 10.5 Shovel swing factor

Angle of swing, in degrees	45	60	75	90	120	150	180
Swing factor	1.2	1.1	1.05	1.00	0.91	0.84	0.77

Source: Atkinson 1992.

TABLE 10.6 Propel time factor

Strip mines	0.75
Open pit mines	0.85
Sand and gravel pits	0.90
High-face quarries	0.95

Source: Atkinson 1992.

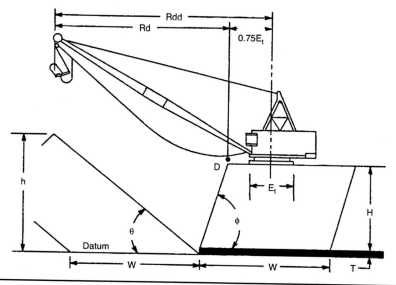

FIGURE 10.7 Dragline pit
Source: Ramani et al. 1980.

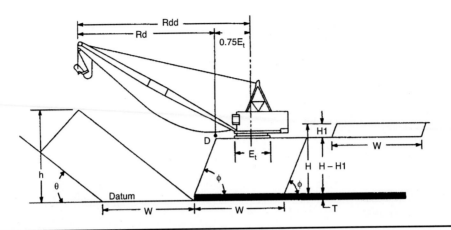

FIGURE 10.8 Dragline pit with a working bench
Source: Ramani et al. 1980.

$$Rdd = Rd + 0.75E_t \qquad \text{(EQ 10.19)}$$

If the dragline works on a bench to increase its operating range (Figure 10.8), and the overburden to depth H1 is removed by some other piece of equipment, the reach equation becomes:

$$Rd = \frac{H - H1}{\tan\phi} + \frac{1}{\tan\theta}\left[(H - H1)\left(1 + \frac{SP}{100}\right) + \left(\frac{W}{4}\tan\theta\right) - T\right] \qquad \text{(EQ 10.20)}$$

SHOVEL SELECTION

The usual operating practice with a stripping shovel is to align one set of its crawlers on the rib of coal adjacent to the spoil. An imaginary plane passing through this location would contain the point D, shown above the front crawler in Figure 10.9. Projection of a horizontal line from the imaginary vertical plane to the point of discharge of the shovel dipper establishes the shovel's reach (Rs), which is calculated as follows:

$$Rs = \frac{1}{\tan\theta}\left[H\left(1 + \frac{SP}{100}\right) + \left(\frac{W}{4}\tan\theta\right) - T\right] \qquad \text{(EQ 10.21)}$$

If Fc is equal to the width across both crawlers (in feet), the shovel's dumping radius becomes:

$$Rsd = Rs + 0.5\,(Fc) \qquad \text{(EQ 10.22)}$$

BULLDOZER SELECTION

Bulldozers are used extensively in surface mining. They are extremely useful for dozing material for short distances under difficult conditions. Selection of a suitable bulldozer can be accomplished by determining the blade size (Bb) in cubic yards. The choice of

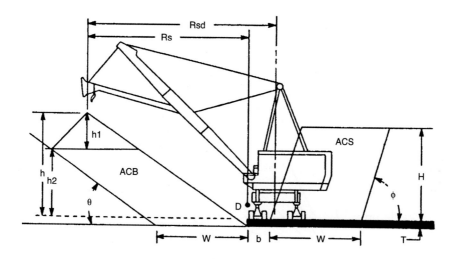

FIGURE 10.9 Shovel pit without a coal fender
Source: Ramani et al. 1980.

blade types depends on job applications and soil conditions. The selection procedures are outlined as follows:

$$FT = MT + LT \qquad \text{(EQ 10.23)}$$

$$TF = \frac{HD}{(S1)(88)} \qquad \text{(EQ 10.24)}$$

$$TE = \frac{HD}{(S2)(88)} \qquad \text{(EQ 10.25)}$$

$$VT = TF + TE \qquad \text{(EQ 10.26)}$$

$$CT = TF + FT \qquad \text{(EQ 10.27)}$$

$$NPH = \frac{60}{CT} \qquad \text{(EQ 10.28)}$$

$$PPH = (NPH)(Bb) \qquad \text{(EQ 10.29)}$$

$$MAXPH = (PPH)(LF)(BF)(AO)(\text{any other applicable correction factor}) \qquad \text{(EQ 10.30)}$$

where FT is the fixed time in minutes; TF is the travel time, loaded, in minutes; TE is the travel time, empty, in minutes; VT is the variable time in minutes; MT is the maneuver time in minutes; CT is the cycle time in minutes; LT is the loading time in minutes; NPH is the number of passes per hour; PPH is the theoretical production per hour in cubic yards; MAXPH is the actual production per hour in cubic yards; HD is the haul distance in feet; S1 is the loaded speed in miles per hour; S2 is the empty speed in miles per hour; Bb is the blade size of the bulldozer in cubic yards; LF is the load factor; BF is the blade factor; and AO is the operating factor. Among the many applicable correction factors available in manufacturers' handbooks are material, grade, visibility, transmission, and dozing factors.

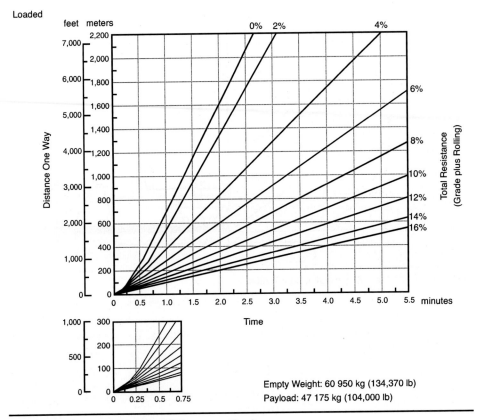

FIGURE 10.10 Loaded scraper (distance vs. time)
Source: Caterpillar 1997.

SCRAPER SELECTION

Scrapers may be elevating or conventional and powered by single or double engines, but the major distinction is between wheeled tractor scrapers and crawler tractor scrapers. Wheel scrapers, because of their mobility, are used in surface coal mining for topsoil removal, road building, etc. Shown in Figures 10.10 and 10.11 are typical scraper cycle times (loaded and empty) as functions of distance and grade.

The following relationships can be established using the same parameters outlined in the section on bulldozers, calculating the cycle time and the number of cycles per hour, and denoting the bucket capacity (BS) in cubic yards:

$$PPH = (NPN)(BS) \qquad \text{(EQ 10.31)}$$

$$MAXPH = (PPH)(LF)(BF)(AO)(\text{any other applicable correction factor}) \qquad \text{(EQ 10.32)}$$

For calculating the number of pusher tractors, the pusher cycle time must be known. The maximum number of scrapers that can be served by one pusher tractor is given by rounding the following ratio to the next lower integer:

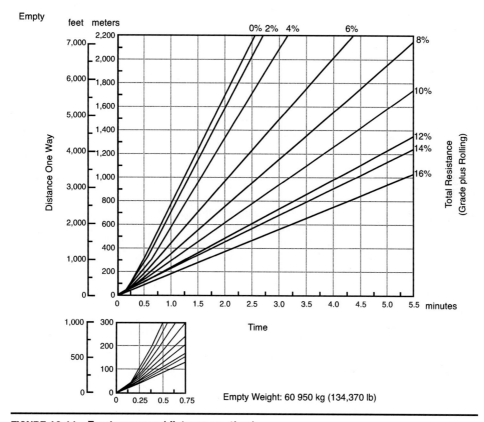

FIGURE 10.11   Empty scraper (distance vs. time)
Source: Caterpillar 1997.

$$\text{Number of scrapers per pusher tractor} = \frac{\text{scraper cycle time}}{\text{pusher cycle time}} \quad \text{(EQ 10.33)}$$

$$\text{Number of pusher tractors} = \frac{\text{total number of scrapers}}{\text{number of scrapers per pusher}} \quad \text{(EQ 10.34)}$$

FRONT-END LOADER SELECTION

In selecting front-end loaders, care must be taken to avoid overestimating the bucket capacity because improper bucket sizing can lead to instability of the equipment. There are three Society of Automotive Engineers (SAE) ratings for front-end loader bucket capacities: (a) struck capacity (that volume contained in a bucket after a load is leveled by drawing a straight edge resting on the cutting edge and the back of the bucket), (b) heaped or rated capacity (struck capacity plus that additional material that would heap on the struck load at a 2:1 angle of repose with the struck line parallel to the ground), and (c) static tipping load (the minimum weight at the center of gravity for the bucket which will rotate the rear of the machine). As in both the bulldozer and scraper calculations, the bucket capacity (BL), in cubic yards, of a front-end loader can be used to determine the actual hourly production in cubic yards.

$$\text{Theoretical production per hour in cubic yards (PPH)} = (BL)(NPH) \quad \textbf{(EQ 10.35)}$$

$$(BL) = \frac{PPH}{NPH} = \frac{MAXPH}{(NPH)(LF)(BF)(AO)(\text{any other applicable correction factor})} \quad \textbf{(EQ 10.36)}$$

where MAXPH is the actual production per hour in cubic yards, NPH is equal to 60/CT, CT is the cycle time in minutes (dumping time in minutes + maneuvering time in minutes + loading time in minutes), LF is the load factor, BF is the bucket factor, and AO is the operating factor.

REFERENCES

Ash, R.L. 1990. Design of Blasting Rounds. In *Surface Mining*. Edited by B.A. Kennedy. Littleton, CO: Society for Mining, Metallurgy, and Exploration, Inc.

Atkinson, T. 1992. Selection and Sizing of Excavation Equipment. In *SME Mining Engineering Handbook*. Littleton, CO: Society for Mining, Metallurgy, and Exploration, Inc.

Austin Powder. 2002. Blasting Formulas. Company Literature.

Caterpillar, Inc. 1997. *Caterpillar Performance Handbook*. Peoria, IL: Caterpillar, Inc.

Caterpillar Tractor Co. 1975. *Fundamentals of Earthmoving*. Peoria, IL: Caterpillar Tractor Company.

Dick, R.A. 1973. Explosives and Borehole Loading. In *SME Mining Engineering Handbook*. New York: Society for Mining, Metallurgy, and Exploration, Inc. and AIME.

Dowding, C.H. and C.T Aimone. 1992. Rock Breakage: Explosives. In *SME Mining Engineering Handbook*. Littleton, CO: Society for Mining, Metallurgy, and Exploration, Inc.

Dyno Nobel. 1997. *Explosives Engineers' Guide*. Salt Lake City, UT: Dyno Nobel.

Hartman, H.L. 1990. Drilling. In *Surface Mining*. Littleton, CO: Society for Mining, Metallurgy, and Exploration, Inc.

Morrell, R.J., and H.F. Unger. 1973. Drilling Machines, Surface. In *SME Mining Engineering Handbook*. New York: Society for Mining, Metallurgy, and Exploration, Inc. and AIME.

Office of Surface Mining. 2001. http://www.osmre.gov.

Pugliese, J.M. 1972. *Designing Blast Patterns Using Empirical Formulas*. I.C. 8550, Washington, DC: US Bureau of Mines.

Ramani, R.V., C.J. Bise, C. Murray, and L.W. Saperstein. 1980. *User's Manual for Premining Planning of Eastern Surface Coal Mining*, vol. 2. Cincinnati, OH: Surface Mine Engineering, US Environmental Protection Agency.

PROBLEMS

Problem 10.1

A dragline is to be selected to remove overburden at a rate of 395,000 bank cubic yards per month. If it will be scheduled to operate 720 hours per month with an operating efficiency of 0.75, its bucket factor is 0.8, and its cycle time is 59 sec, what should be the size of its bucket?

Problem 10.2

Management personnel at a surface mine needs to verify that the blasting rounds conform to federal ground-vibration regulations. As shown in the figure below, a house that needs to be protected is located 1,500 ft due west of a proposed blast and 500 ft beyond the permit boundary.

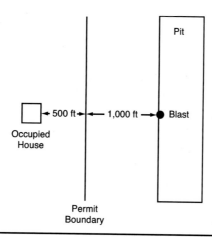

Illustrates Problem 10.2

The mine has an approved attenuation formula for the site as a result of data taken from 50 monitored blasts at the site:

$$V = 242.75(Ds)^{-1.70}$$

Determine the "modified scaled distance factor" and the maximum explosive charge weight per 8-m-sec delay period that are necessary to protect the home.

Problem 10.3

Suppose that the surface mining company mentioned in problem 10.2 did not have the necessary data for the Office of Surface Mining to approve the attenuation formula. How would that change your answer to problem 10.2?

Problem 10.4

Specify the blasting-pattern dimensions for a surface coal mine. The overburden strata are flat with little or no surface soil to consider in the blasting design. The generalized section of the overburden shows 42 ft of sandstone and shale overlying a 4-ft coal seam. A 6-in. drill is on site. For cost considerations, bulk-load ammonium nitrate/fuel oil should be used, if possible.

Problem 10.5

How many drilling units are required to meet the overburden-preparation requirements of a surface mine if the following information is provided?

Spacing of the drill holes: 22 ft
Burden of the drill holes: 12 ft
Cubic yards of overburden to be removed per month: 395,000
Hours scheduled per shift: 6
Shifts per day: 2
Drilling days per week: 5
Weeks worked per month: 4
Drilling rate: 80 feet per hour

Problem 10.6

Based on the information provided in the previous problem, determine the amount of explosives required on a monthly basis if the powder factor is 1.0.

Problem 10.7

Determine the maximum height of highwall that a dragline can strip without rehandling in level terrain, using the following information:

Dumping radius of the dragline: 154 ft
Tub diameter: 36 ft
Dragline is offset 0.75 (tub diameter)
Spoil's angle of repose: 37°
Highwall's angle of repose: 74°
Pit width: 50 ft
Coal thickness: 4 ft
Swell: 30%

Problem 10.8

Given the same conditions as in problem 10.7, if another form of stripping equipment removes the top 8 ft of the overburden, determine the new maximum highwall height.

Problem 10.9

A coal company is considering the recovery of a 6-ft seam under an average of 100 ft of overburden using a dragline. The dragline will use a straight side-casting method. A contract to be signed requires 1.25 million tons of coal to be mined annually.

Given the following information, determine the reach requirements and minimum bucket capacity for the dragline.

Coal weight: 0.04 tons per cu ft
Recovery: 90%
Overburden swell: 35%
Overburden density (in place): 3,400 lb per cu yd
Highwall angle (ϕ): 71.6°
Spoil angle of repose (θ): 38.7°
Pit width: 100 ft
Average dragline cycle time: 55 seconds per cycle
Overall dragline efficiency: 0.81
Bucket factor: 0.90
Dragline schedule: 720 hours per month

Problem 10.10

A small deep-mining company has decided to recover some of its strippable reserves. Instead of investing large amounts of capital into the venture, the decision was made to purchase used equipment. The company located a secondhand stripping shovel with a 25-cu-yd bucket. With this shovel, the company intends to strip the overburden from a 4-ft coal seam that outcrops with a 2-ft cover and then extends under a gentle hillside that has a slope of 1 in 12. The shovel will be scheduled 336 hr per month. Given the following information, calculate the number of strip pits the company can develop and the strip ratio and monthly production of coal in the last pit.

> Bucket factor: 0.85
> Cycle time: 50 sec
> Monthly operating factor: 0.70
> Overburden swell: 0.30
> Pit width: 60 ft
> Dumping radius of the shovel: 100.25 ft
> Highwall angle of repose: 74°
> Spoil angle of repose: 37°
> Width across the shovel's crawlers: 38.75 ft
> Coal weight: 80 lb per cu ft

Problem 10.11

An interesting relationship known as the maximum usefulness factor (MUF), which relates a stripping machine's weight and its ability to do work, was developed during the early 1960s. In actuality, the MUF is equal to the product of the nominal bucket size, in cubic yards, multiplied by the machine's functional dumping reach, in feet. In this manner, two similar machines with different reaches or bucket sizes can be related.

If the dumping radius of an 18-cu-yd dragline is 154 ft, the tub diameter of the machine is 36 ft, and normal operation requires the machine to be offset a distance of three quarters of the tub diameter from the edge of the highwall. Determine the potential effective reach of the machine if it is to be refitted with a new boom and a 15-cu-yd bucket.

Problem 10.12

The owner of the company mentioned in problem 10.10 decided that a secondhand dragline might be a better choice than the stripping shovel. A dragline identical to the one described in problem 10.11 was located. If the dragline operation is to differ from the shovel operation only in terms of bucket factor (0.80), cycle time (59 sec), and monthly operating factor (0.75), determine the maximum number of pits that the dragline can develop and the strip ratio and monthly production in the last pit.

Problem 10.13

Using the following production and correction charts supplied by a bulldozer manufacturer, determine the actual production per hour (in cubic yards) for a D9R-9U machine moving loose stockpile material (2,500 lb per loose cubic yard) 150 ft down a 10% grade using a slot-dozing technique. The operator is considered average, and his job efficiency is estimated to be 50 min per hr.

JOB CONDITION CORRECTION FACTORS

	TRACK-TYPE TRACTOR	WHEEL-TYPE TRACTOR
OPERATOR —		
Excellent	1.00	1.00
Average	0.75	0.60
Poor	0.60	0.50
MATERIAL —		
Loose stockpile	1.20	1.20
Hard to cut; frozen —		
with tilt cylinder	0.80	0.75
without tilt cylinder	0.70	—
cable controlled blade	0.60	—
Hard to drift; "dead" (dry, non-cohesive material) or very sticky material	0.80	0.80
Rock, ripped or blasted	0.60-0.80	—
SLOT DOZING	1.20	1.20
SIDE BY SIDE DOZING	1.15-1.25	1.15-1.25
VISIBILITY —		
Dust, rain, snow, fog or darkness	0.80	0.70
JOB EFFICIENCY —		
50 min/hr	0.83	0.83
40 min/hr	0.67	0.67
BULLDOZER*		
Adjust based on SAE capacity relative to the base blade used in the Estimated Dozing Production graphs.		
GRADES — See following graph.		

*Note: Angling blades and cushion blades are not considered production dozing tools. Depending on job conditions, the A-blade and C-blade will average 50-75% of straight blade production.

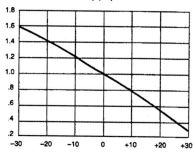

% Grade vs. Dozing Factor
(−) Downhill
(+) Uphill

ESTIMATING DOZER PRODUCTION OFF-THE-JOB

Example problem:

Determine average hourly production of a D8R/8SU (with tilt cylinder) moving hard-packed clay an average distance of 45 m (150 feet) down a 15% grade, using a slot dozing technique.

Estimated material weight is 1600 kg/Lm³ (2650 lb/LCY). Operator is average. Job efficiency is estimated at 50 min/hr.

Uncorrected Maximum Production — 458 Lm³/h (600 LCY/hr) (example only)

Applicable Correction Factors:
Hard-packed clay is "hard to cut" material –0.80
Grade correction (from graph) ....–1.30
Slot dozing ....–1.20
Average operator ....–0.75
Job efficiency (50 min/hr) ....–0.83
Weight correction ....(2300/2650)–0.87

Production = Maximum Production × Correction Factors
= (600 LCY/hr) (0.80) (1.30) (1.20) (0.75) (0.83) (0.87)
= 405.5 LCY/hr

To obtain production in metric units, the same procedure is used substituting maximum uncorrected production in Lm³.
= 458 Lm³/h × Factors
= 309.6 Lm³/h

Illustrates Problem 10.13

Problem 10.14

For a 30.6-bank-cu-yd capacity scraper, whose loaded and empty tram times are shown in the following figures, estimate the production and cycle time carrying its rated capacity. The 5,000-ft-long level haul road is firm, smooth, and maintained fairly regularly. Load time is assumed to be 0.65 min, and the dump and maneuver time is estimated to be 0.75 min. Assume 90% job efficiency.

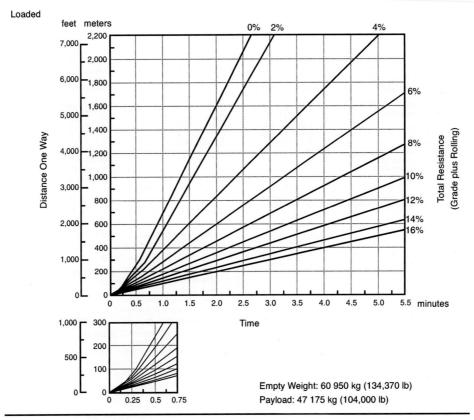

Illustrates Problem 10.14

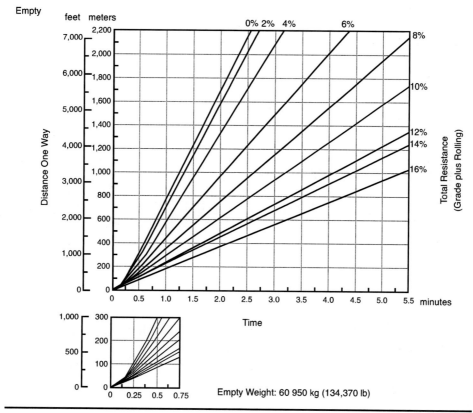

Illustrates Problem 10.14

Problem 10.15

A surface coal mine operator needs to purchase a front-end loader that will be used to load ripped coal onto on-highway dump trucks. The mine operates two 7-hour shifts per day, 5 days per week, 50 weeks per year. The daily production of coal is 1,500 tons. The desired cycle time is 1.0 min or less. The density of loose coal is approximately 1,400 lb per cu yd. Assume that the load factor is 0.65, the bucket factor is 1.23, and the operating factor is 0.75. What bucket size is required?

PROBLEM SOLUTIONS

Solution 10.1

The number of theoretical cycles per month:

$$\text{NTCPM} = \frac{(\text{SMH})(3{,}600)}{\text{CT}} = \frac{(720)(3{,}600)}{59} = 43{,}932$$

Actual cycles per month:

$$\text{ACM} = \text{NTCPM}(\text{OF}) = 43{,}932(0.75) = 32{,}949$$

Bank cubic yards to be handled per cycle:

$$\text{BCY} = \frac{\text{monthly requirement}}{\text{ACM}} = \frac{395{,}000}{32{,}949} = 12 \text{ cu yd}$$

Bucket capacity (loose volume):

$$\text{BC}_{\text{loose}} = \frac{\text{BCY}}{\text{BF}} = \frac{12}{0.8} = 15 \text{ cu yd}$$

Solution 10.2

According to 30 CFR (Code of Federal Regulations) 816.67(d), the maximum allowable peak particle velocity for ground vibration (V) for a structure to be protected that is located 1,500 ft from a blast is 1.00 in. per sec. If you take the approved attenuation formula for the site and make the above substitution, you get:

$$V = 242.75(\text{Ds})^{-1.70}$$

$$(1.00) = 242.75(\text{Ds})^{-1.70}$$

$$\text{Ds} = 25.54$$

Next, by rearranging Eq. 10.2, the charge weight per delay becomes:

$$W = \left(\frac{D}{D_s}\right)^2$$

where:

W = maximum weight of explosives per 8-m-sec delay (in pounds)

D = distance from the home (in feet)

Thus,

$$W = \left(\frac{D}{D_s}\right)^2 = \left(\frac{1{,}500}{25.54}\right)^2 = 3{,}450 \text{ lb}$$

Solution 10.3

According to 30 CFR (Code of Federal Regulations) 816.67(d), the scaled distance factor (D_s) necessary in the absence of data for a structure to be protected that is located 1,500 ft from a blast is 55. Therefore, the charge weight per delay becomes:

$$D = (D_s)(\sqrt{W})$$

$$1{,}500 = (55)(\sqrt{W})$$

$$27.27 = (\sqrt{W})$$

$$744 \text{ lb} = W$$

Solution 10.4

Utilizing the empirical relationships presented in this chapter, the first step is to determine the burden (B):

$$B \text{ (ft)} = \frac{K_B D_E}{12} = \frac{25(6 \text{ in.})}{12 \text{ in. per ft}} = 12.5 \text{ ft}$$

A practical layout in the field is 12.0 ft.

Next, determine the spacing (S):

$$S \text{ (ft)} = K_S B = 1.8(12) = 21.6 \text{ ft}$$

To keep K_S within the range of 1.8 to 2.0, use S = 22 ft.

Next, determine the subdrilling (J). Since there is no subdrilling when benching to coal (the powder column is terminated 3 to 5 ft above the coal), a –J of 5 ft is used.

Next, determine the hole-length ratio (K_H):

$$K_H = \frac{H}{B} = \frac{L + (-J)}{12} = \frac{42 + (-5)}{12} = 3.08$$

Because K_H is greater than 1 and less than 4, this is acceptable. However, if it is not performing well, thought should be given to raising the ammonium nitrate/fuel oil density, increasing the drilling diameter, or changing the mining plan.

Finally, determine the stemming (T):

$$T = K_T B = (0.7)(12) = 8.4 \text{ ft}$$

For practicality, use a stemming of 9 ft.

A typical layout for the situation analyzed here would be plan C in Figure 10.6.

Solution 10.5

$$\text{Length of drilling in feet per month} = \frac{COB}{DP}(27)$$

$$= \left(\frac{395{,}000}{(22)(12)}\right)(27)$$

$$= 40{,}398 \text{ ft}$$

Scheduled monthly hours = (HPS)(SPD)(DPW)(SPM)

$$= (6)(2)(5)(4)$$

$$= 240 \text{ hr}$$

$$\text{Number of drills required} = \frac{LDPM}{(SMH)(DR)}$$

$$= \frac{40{,}398}{(240)(80)}$$

$$= 2.1 \text{ drills}$$

Thus, two drills would probably be used, and any additional required drilling will be accomplished during an extra shift or on weekends.

Solution 10.6

$$\text{Amount of explosives, in pounds, required per month} = \frac{COB}{PK}$$

$$= \frac{395{,}000}{1.0}$$

$$= 395{,}000 \text{ lb}$$

Solution 10.7

Using Eq. 10.19 and 10.18:

$$Rdd = Rd + 0.75(E_t)$$

$$154 = Rd + 0.75(36)$$

$$Rd = 127 \text{ ft}$$

$$Rd = \frac{H}{\tan\phi} + \frac{1}{\tan\theta}\left[H\left(1 + \frac{SP}{100}\right) + \left(\frac{W}{4}\tan\theta\right) - T\right]$$

$$127 = \left(\frac{H}{3.49}\right) + \left(\frac{1}{0.754}\right)\left[H(1.3) + \left(\frac{50}{4}\right)(0.754) - 4\right]$$

$$127 = 0.287(H) + 1.326(1.3H + 9.425 - 4)$$

$$119.8 = 2.01H$$

Thus, the maximum highwall height is 59.6 ft.

Solution 10.8

Using Eq. 10.20:

$$Rd = \left(\frac{H-H1}{\tan\phi}\right) + \frac{1}{\tan\theta}\left[(H-H1)\left(1+\frac{SP}{100}\right) + \left(\frac{W}{4}\right)\tan\theta - T\right]$$

$$127 = \left(\frac{H-8}{3.49}\right) + \frac{1}{0.754}\left[(H-8)(1.3) + \left(\frac{50}{4}\right)(0.754) - 4\right]$$

$$127 = (0.287)(H) - (2.29) + (1.724)(H) - (13.79) + (12.5) - (5.304)$$

$$127 = 2.01H - 8.88$$

$$135.88 = 2.01H$$

Thus, the maximum highwall height is now 67.6 ft.

Solution 10.9

From Eq. 10.18, the reach of the dragline (Rd) is:

$$Rd = \left(\frac{H}{\tan\phi}\right) + \frac{1}{\tan\theta}\left[(H)\left(1+\frac{SP}{100}\right) + \left(\frac{W}{4}\right)\tan\theta - T\right]$$

$$= \frac{100}{3} + \frac{1}{0.8}\left[(100)\left(1+\frac{35}{100}\right) + \left(\frac{100}{4}\right)(0.8) - 6\right]$$

$$= 33.33 + (1.25)[(135) + (20) - 6]$$

$$= 220 \text{ ft}$$

In a 6-ft seam, 1 sq ft of exposed coal is equal to 0.24 tons of coal; therefore, to produce 1.25 million tons annually, 5,200,000 sq ft of coal must be exposed, which is equivalent to 578,700 sq yd.

With an average of 100 ft of cover (equivalent to 33.33 yd), a total of 19,265,000 cu yd must be moved annually. This is equivalent to 1,605,395 yd per month. Assuming 90% recovery, the necessary monthly yardage is:

$$1,605,395/0.9 = 1,783,772 \text{ cu yd per month}$$

Because the dragline is scheduled 720 hours per month, the necessary hourly yardage (Q) is 1,783,772/720 = 2,478 cu yd. Now, to determine the necessary bucket size, refer to Eq. 10.17.

$$B_c = \frac{Q}{(C)(S)(A)(O)(B_f)(P)}$$

$$C = \text{theoretical cycles per hour} = \frac{3,600 \text{ seconds per hour}}{55 \text{ seconds per cycle}} = 65.45$$

$$S = \text{swing factor} = 1.00 \text{ (from Table 10.4)}$$

$$(A)(O)(P) = 0.81$$

$$B_f = 0.90$$

$$B_c = \frac{2{,}478}{(65.45)(1.00)(0.81)(0.90)}$$

$$= 51.9 \text{ cu yd (bank volume)}$$

When swell is taken into account, the bucket size based on loose volume is $(51.9)(1.35) = 70$ cu yd.

Solution 10.10

The first step in solving this type of problem is to determine the bank cubic yards that the shovel can move in 1 month:

Bucket size in cubic yards: 25
Bucket factor: 0.85
Bucket load in bank cubic yards: $(25)(0.85) = 21.25$
Cycle time in seconds: 50
Passes per hour: (3,600 sec per hour)/(50 sec) = 72
Theoretical bank cubic yards per hour: $(72)/(21.25) = 1{,}530$
Monthly operating hours: 336
Theoretical bank cubic yards per month: 514,080
Monthly operating factor: 0.70
Bank cubic yards per month: $(514{,}080)(0.70) = 359{,}856$

The maximum highwall height can be determined using Eq. 10.21 and 10.22:

$$Rs = Rsd - [0.5(Fc)] = 100.25 - [0.5(38.75)] = 81.0$$

$$Rs = \frac{1}{\tan\theta}\left[H\left(1 + \frac{SP}{100}\right) + \left(\frac{W}{4}\right)\tan\theta - T\right]$$

$$81.0 = \frac{1}{0.754}\left[H(1.3) + \left(\frac{60}{4}\right)0.754 - 4\right]$$

$$81.0 = (1.326)[(1.3H) + (11.31) - (4)]$$

$$81.0 = (1.326)[(1.3H) + (7.31)]$$

$$81.0 = (1.72H) + (9.69)$$

$$71.3 = 1.72H$$

$$H = 41.5 \text{ ft}$$

Therefore, H is 41.5 ft. To determine the number of strip pits that the company can develop, it must be realized that the shovel is constrained by its ability to spoil into the previously mined pit. Thus, under gradually increasing overburden, the average height of the overburden for a particular pit must not exceed 41.5 ft. The number of pits can be determined as follows:

Pit Number	Average Overburden Depth	Is it minable?
1	(2 + 7)/2 = 4.5	Yes
2	(7 + 12)/2 = 9.5	Yes
3	(12 + 17)/2 = 14.5	Yes
4	(17 + 22)/2 = 19.5	Yes
5	(22 + 27)/2 = 24.5	Yes
6	(27 + 32)/2 = 29.5	Yes
7	(32 + 37)/2 = 34.5	Yes
8	(37 + 42)/2 = 39.5	Yes
9	(42 + 47)/2 = 44.5	No

Therefore, eight pits can be developed.

In the eighth pit, each linear foot of advance requires the removal of the following bank cubic yards:

$$\frac{(39.5 \text{ ft})(60 \text{ ft})}{27 \text{ cu ft per cu yd}} = 87.8 \text{ cu yd per ft of advance}$$

Through this same distance, the following coal tonnage is exposed:

$$\frac{(4 \text{ ft})(60 \text{ ft})(80 \text{ lb per cu ft})}{2,000 \text{ lb per ton}} = 9.6 \text{ tons}$$

Because the stripping ratio is defined as the yardage of overburden divided by the tonnage of coal exposed, it is as follows for this situation:

$$\frac{87.8 \text{ cu yd}}{9.6 \text{ tons}} = 9.2 \text{ cu yd per ton}$$

The monthly production in the eighth pit is as follows:

$$\frac{359,856 \text{ cu yd}}{9.2 \text{ cu yd per ton}} = 39,115 \text{ tons of coal}$$

Solution 10.11

Rdd for the 18-cu yd machine = 154 ft

Rd for the 18-cu yd machine = 154 − [(0.75)(36)]

= 154 − 27 = 127 ft

$(Rd_{18})(18 \text{ cu yd}) = (Rd_{15})(15 \text{ cu yd})$

Thus,

$$\text{Rd for the 15-cu yd machine} = \frac{127(18)}{15} = 152.4 \text{ ft}$$

Solution 10.12

The bank cubic yards handled per month by the dragline are as follows:

Bucket size: 15
Bucket factor: 0.80
Bucket load in bank cubic yards: 12
Cycle time in seconds: 59
Passes per hour: 61
Theoretical bank cubic yards per hour: 732
Monthly operating hours: 336
Theoretical bank cubic yards per month: 245,952
Monthly operating factor: 0.75
Bank cubic yards per month: 184,464

The dragline's reach is as follows:

$$Rd = \frac{H}{\tan\phi} + \frac{1}{\tan\theta}\left[H\left(1 + \frac{SP}{100}\right) + \left(\frac{W}{4}\right)\tan\theta - T\right]$$

$$152.4 = \frac{H}{\tan 74°} + \frac{1}{\tan 37°}\left[H(1.3) + \left(\frac{60}{4}\right)\tan 37° - 4\right]$$

$$152.4 = (0.287H) + (1.33)(1.3H + 11.3 - 4)$$

$$152.4 = (0.287H) + [(1.73H + 9.71)]$$

$$142.7 = 2.02H$$

$$H = 70.64$$

Therefore, H is 71 ft.

The number of strip pits that the company can now develop is as follows:

Pit Number	Average Overburden Depth	Is it minable?
10	(47 + 52)/2 = 39.5	Yes
11	(52 + 57)/2 = 54.5	Yes
12	(57 + 62)/2 = 59.5	Yes
13	(62 + 67)/2 = 64.5	Yes
14	(67 + 72)/2 = 69.5	Yes
15	(72 + 77)/2 = 74.5	No

Therefore, 14 pits are now possible.

Strip ratio in the fourteenth pit:

$$\frac{(69.5 \text{ ft})(60 \text{ ft})}{(27 \text{ cu ft per cu yd})(9.6 \text{ tons})} = 16.1 \text{ cu yd per ton}$$

Monthly production in the fourteenth pit:

$$\frac{184,464 \text{ cu yd}}{16.1 \text{ cu yd per ton}} = 11,457 \text{ tons of coal}$$

SURFACE EXTRACTION | 303

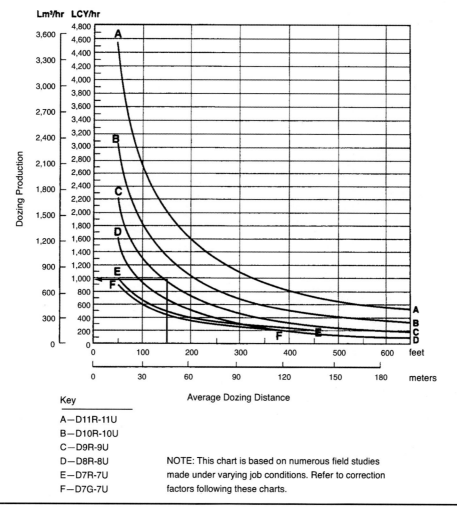

Illustrates Solution to Problem 10.13

Solution 10.13

From the production chart provided, the uncorrected maximum production is 950 loose cu yd per hr.

The applicable correction factors are as follows:

    Loose stockpile material (from table): 1.20
    Grade correction (from diagram): 1.2
    Slot dozing (from table): 1.20
    Average operator (from table): 0.75
    Job efficiency (from table): 0.83

Weight correction: The bulldozer production curves are based on a soil density of 2,300 lb per loose cubic yard. Since the material handled in this problem weighs 2,500 lb per loose cubic yard, a weight correction of 0.92 must be incorporated.

MINING ENGINEERING ANALYSIS

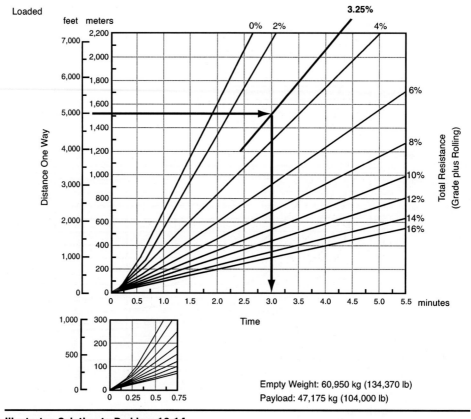

Illustrates Solution to Problem 10.14

$$\text{Actual hourly production} = (\text{maximum hourly production})(\text{correction factors})$$
$$= 950(1.20)(1.2)(1.20)(0.75)(0.83)(0.92)$$
$$= 940 \text{ loose cu yd per hr}$$

Solution 10.14

Since the capacity of the machine is already given in bank cubic yards, all that needs to be determined is the cycle time.

Grade: 0% for a level haul road
Rolling resistance: 65 lb per ton = 3.25% (see Table 9.1)
Total effective grade: 3.25%

Cycle time
Load time:	0.65 min
Dump and maneuver time:	0.75 min
Haul time (5,000 ft):	3.00 min (see figure)
Return time (5,000 ft):	<u>2.10 min</u> (see figure)
TOTAL:	6.50 min

SURFACE EXTRACTION | 305

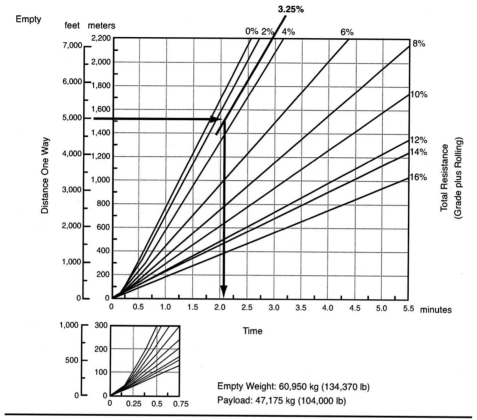

Illustrates Solution to Problem 10.14

$$\frac{60 \text{ min per hr}}{(6.5 \text{ min per cycle})(0.90)} = 8.31 \text{ cycles per hr}$$

(8.31 cycles per hr)(30.6 bank cu yd) = 254 bank cu yd per hr

Solution 10.15

Recalling that NPH = 60/CT, CT = DP + MS + LT, and that it is desired to maintain CT equal to or less than 1.0 min, the worst case situation would occur when CT = 1.0 min. Thus, use NPH = 60.

The required MAXPH is:

$$\text{MAXPH} = \left(\frac{1,500 \text{ tons}}{\text{day}}\right)\left(\frac{1 \text{ day}}{14\text{hr}}\right)\left(\frac{2,000 \text{ lb}}{1 \text{ ton}}\right)\left(\frac{1 \text{ cu yd coal}}{1,400 \text{ lb}}\right)$$

= 153 cu yd per hr

From Eq. 10.36:

$$(BL) = \frac{PPH}{NPH}$$

$$= \left\{\frac{MAXPH}{(NPH)[(LF)(BF)(AO)(\text{any other applicable correction factor})]}\right\} \frac{153}{60[(0.65)(1.23)(0.75)]}$$

= 4.25 cu yd

Thus, select a machine with a bucket in excess of 4.25 cu yd.

Index

Note: *f* indicates figure; *t* indicates table.

A

a-c loads   153, 154
Absolute discharge   149
Absolute intake pressure   149
Acceleration   6
   and displacement   6
   and velocity   6
Adiabatic compressed-air process   148*f*, 149
Advance mining   4
Air courses   4
Air splits   4
Airborne contaminants   28
Airways and airflow. *See also* Ventilation
   airway defined   4
   ample capacity   51
   friction factors   52, 53*t*, 54*t*
   principles   52, 52*t*
American Conference of Governmental Industrial Hygienists (ACGIH)   28
Ampacity   155, 156*t*, 157*t*
Area-of-influence method. *See* Polygonal method
Arithmetic mean   8
   and control charts   32–33
Available power   245–246, 246*f*
Average, defined   8

B

Bank density   281, 282*t*
Barrier pillars   4
Beam building   4
Belt conveyors   215, 221
   arrangements   221, 224*f*
   belt speed   221, 222*t*, 223*t*
   belt tension   221–225
   belt weight (B) and weight of moving parts (Q), 222–225, 225*t*
   belt width   221, 222*t*
   components   221
   drive factor   224, 226*t*
   friction factors   222, 225*t*
   idler spacing   225, 226*t*
   lump size   221, 222*t*
   motor horsepower   226
   normal bulk material capacity for belt speeds   223*t*
Belt idlers   4
Belt take-up   4
Bernoulli theorem   8
Bieniawski formula   85–86
Blasting   271, 272. *See also* Drilling
   borehole pressure   272
   density   272
   detonation pressure   272, 273*f*
   detonation velocity   272
   explosive types   272–273
   federal blast vibration standards   274, 275*f*
   and fragmentation procedures   272
   loading density (LD)   272, 273*f*
   powder factor (amount of explosives required)   280
   specific gravity   272
   water density   272
Blasting-round design   273, 274*f*
   bench height (L)   274
   burden (B)   274, 277, 279
   burden ratio   276
   calculations and worksheet   277–279, 278*t*
   collar distance ratio and calculation   276, 278
   collar length (T)   274
   diameter (D_e)   274
   empirical approach   273, 274–275, 275*f*, 276*f*
   generalized patterns   274, 276*f*
   ground-vibration approach   273–274, 275*f*
   hole length ratio and calculations   276, 278
   length of hole (H)   275
   ratios (K)   275–277
   separation between adjacent holes (s)   274
   single-hole dimensions (abbreviations)   274, 275*f*
   spacing (S)   274, 277, 279
   spacing between rows of holes (b)   274
   spacing ratio   276
   subdrilling length (J)   274, 278, 279
   subdrilling ratio   276
Bleeder entries   4
Bourdon gages   121
Boyle's law   7–8
Breakthrough   4
British thermal units (Btu)   4
Bucket capacity   280
   front-end loaders   287–288
   parameters   281*t*
   scrapers   286

Bulldozers
    blade size (Bb) 284–285
    selection procedure and factors 285
Burden
    calculating 277, 279
    defined 4
    ratio 276

C

Cages 5
Capacitive reactance (C) 154–155
Cars 5
Charles's law 7–8
Clean Air Amendments Act 21–22
CMRR. *See under* Coal mining
Coal mining
    airborne contaminant regulations 28
    coal mine roof rating (CMRR) 87
    coal strength and other properties 79, 81t
    compliance requirements 21–22
    friction factors for airways and openings 52, 54t
    MSHA safety and health regulations 28, 29t, 30t
    pillar strength and specimen size 84–85
    roof spans 87
    subsidence control 88–94, 90f, 91f, 91t, 92f, 92t, 93f
    typical staffing for room-and-pillar and longwall mines 24, 24t–26t
Collar
    defined 5
    distance calculation 276, 278, 279
    distance ratio 276
    length (T) 274
Compressed-air power 145
    absolute discharge 149
    absolute intake pressure 149
    absolute pressure 146
    adiabatic process 148f, 149
    air receivers 145
    air requirements of representative drilling machines 151t, 152
    atmospheric pressure at different altitudes 150–151, 151t
    barometric pressure 146
    compression ratio 149
    compression start 146, 147f
    compression stroke 146, 147f
    cylinder clearance 146, 147f, 149, 149f
    delivery stroke 146, 147f
    discharge pressure 145
    discharge valves 146
    distribution system 145
    double-acting pistons 146
    expansion stroke 146–147, 147f
    factors for correcting actual capacity of compressors when used at altitudes above sea level 150, 150t
    gage pressure 146
    indicated horsepower 149
    intake (suction) stroke 147, 147f
    intake valves 146
    isothermal process 148f, 149
    line loss (pressure loss in hose) 152, 152t–153t
    multipliers for air capacity to operate multiple rock drills 150, 150t
    multipliers for altitudes above sea level 149–150, 150t
    reciprocating compressors 145, 146–147
    single-acting pistons 146
    single-stage compressors 146–147, 147f
    system design 149–152, 150t, 151t, 152t, 153t
    theory and equation 148–149, 148f
    two-stage compressors (and theoretical indicator card) 147, 148f
Compressive stress 79
Control charts 32–33, 32f, 33f
    distribution range 32–33, 33f
    lower control limit 32
    upper control limit 32
Conservation of energy 7
Cover 5
CPM (critical path method) 26
Crosscuts 5
Current (I) 153, 155

D

d-c loads 153
Development 5
Displacement 5
    and acceleration 6
    and velocity 6
Draglines and dragline pits 280, 283f
    reach 281–284, 284f
Drainage 121. *See also* Pipes, Pumps and pumping
Drawbar pull 7
Drilling 271. *See also* Blasting, Blasting-round design
    attack 271
    calculating number of drills required 279
    drag bits 272
    percussion 271
    rolling cutter bits 272
    rotary 271
Drum hoists 197, 198f
    horsepower calculations (root mean square method) 200–202
    horsepower variables 198–200
Dump, defined 5

E

Effective grade (total resistance) 245
Efficiency 6
Electrical power 153–155
    a-c loads 153, 154
    ampacity 155, 156t, 157t
    capacitive reactance (C) 154–155
    in continuous-miner systems 154–155, 154f
    correction factors for cables with one or more layers wound on a reel 156, 157t
    correction factors for insulations at various ambient temperatures 155, 156t

current (I)   153, 155
d-c loads   153
default data tables   162, 163t
demand   162
demand factor   162
impedance (Z)   154–155
inby current   162
inductive reactance (L)   154, 155, 158t, 159t
load factor   162
load-flow analysis   153, 162
mine load center   162
outby current   162
phase angle   154
power (P)   153
power billing   162–163, 164t–165t
power cable selection   155–158, 156t, 157t, 158t
power factor (PF)   153, 158–160, 160t
power-factor correction   158, 160–161, 161f, 161t
power-factor-correction capacitors   153
power-factor triangles   160–161, 161f
rectifiers   153
resistance (R)   154, 155, 158t, 159t
surge capacitors   153
system components   153
system schematic   153–154, 154f
tangent (kVAR) table   160, 161t
transformers   153
transmission lines   153
voltage (V)   153
voltage drop   155–158
Entries   5
Equivalent effective weight (EEW)   198, 200f
Excavation   1, 2f, 271
  bank density   281, 282t
  bucket capacity   280, 281t, 286, 287–288
  bulldozer selection   284–285
  draglines and dragline pits   280–284, 283f, 284f
  fillability   281, 282t
  front-end loader selection   287–288
  propel time factor   281, 283t
  scraper selection   286–287, 286f
  shovel operating efficiency   280, 281t
  shovel selection   284
  shovel swing factor   281, 283t
  swell factor   281, 282t
Excavation and handling system. *See* Excavation, Haulage systems

F

Face haulage. *See* Underground face-hauling vehicles
Faces   5
Fans. *See* Ventilation
Fillability   281, 282t
Force   6
Friction hoists. *See* Koepe hoists
Front-end loaders
  calculating production   287–288
  Society of Automotive Engineers ratings   287

G

Gas laws   7–8
Gay-Lussac's law   7–8
Gob, defined   5
Grade resistance (assistance)   7
  rail haulage   218
  trucks   245
Gross vehicle weight (GVW)   245

H

Haulage systems   1, 2f *See also* Belt conveyors, Locomotives, Rail haulage, Trucks, Underground face-hauling vehicles
Head
  components   123, 125f
  defined   5
  frictional head   123, 124t
  static   125–126
  total   122–123
  velocity   123
Headframes   5
Hoists   195, 197. *See also* Wire ropes
  drum dimensions   198, 199f
  drum hoist horsepower calculations (root mean square method)   200–202
  drum vs. Koepe   197, 198f
  equivalent effective weight (EEW)   198, 200f
  horsepower variables (drum hoists)   198–200
  Koepe hoists horsepower calculations (root mean square method)   202–204, 203f
  net weight of the loads   197
  rope size and weight   197–198, 199f
  shaft/slope parameters   197–198
  skip load (SL)   197
  skip weight (SW)   197
Hole length
  calculations   278
  ratio   276, 278
Holland-Gaddy equation   85
Horsepower. *See under* Belt conveyors, Compressed-air power, Hoists, Locomotives, Trucks
Hyperbolic tangent function   89

I

Impedance (Z)   154–155
Inby, defined   5
Incidence rate (IR)   28
Indicated horsepower   149
Inductive reactance (L)   154, 155, 158t, 159t
Intake   5
Isothermal compressed-air process   148f, 149

K

Kinetic energy   7
Kirchhoff's laws   7
Koepe hoists   197, 198f
  horsepower calculations (root mean square method)   202–204, 203f

L

Lateral stress   79

Line loss
  pressure loss in compressed-air hose 152, 152t–153t
  voltage drop in power cable 155–158
Locomotives. *See also* Rail haulage
  acceleration 218–219
  adhesive values 217, 218t
  curve resistance 218
  deceleration 218–219
  drawbar pull 217
  grade resistance 218
  horsepower 217–218
  motor duty cycle (root mean square method) 219–220
  rolling resistance 218
  tractive effort 7, 217–218, 218f
  tractive effort required for hauling on grade against loads 219
  tractive effort required for hauling when grades are in favor of loads 219
Longwall mining, shield-type roof supports and capacity 88, 89f
Lower control limit 32

M

Main entries 5
Maximum (static) rope pull (RPS) 195–197
Mean 8
  and control charts 32–33
Mine planning 1–2, 2f
  and causes of variation in mining 8
Mine preplanning 1, 2f, 19
  mine safety and health 27–31, 29t, 30t, 31t
  production planning 22–23
  project scheduling 26–27
  and quality control 31–33, 32f, 33f
  reserve estimation 19–22, 20f, 20t, 21f, 22f
  staff planning 24, 24t–26t
Mine Safety and Health Administration (MSHA) 27–28
  coal mine regulations 28, 29t, 30t
  incidence rate 28
  indices 27–28
  and noise control 29–31
  severity measure 28
Modulus of elasticity 7

N

National Coal Board (NCB), United Kingdom 90
Noise control 28–29
  combining decibel levels 29–30, 31t
  permissible exposures 31, 31t
Noncoal mining, pillar strength and specimen size 85

O

Ohm's law 7
Openings
  boundary stress concentrations around 79–80, 82f, 83f
  compressive stress concentrations around openings of various shapes 80, 84f
  friction factors 52, 54t
  stresses around 79–80, 80f
  width for shallow overburden 86
Operating support system 1, 2f
Outby, defined 5

P

Panels 5
Pennsylvania Bituminous Mine Subsidence Act of 1966 93–94, 93f
PERT (program evaluation and review technique) 26–27, 27f
Phase angle 154
Pillars 82
  Holland–Gaddy equation 85
  strength equations 84–86
  tributary area 82
Pipes
  characteristic curves 122–123
  equivalent number of feet of straight pipe for different fittings 123t
  frictional head 123, 124t
  head components 123, 125f
  operating point 123
  point of intersection 123
  total head 122–123
  velocity head 123
Pit haulage. *See* Trucks
Planar failure 94–95, 94f
Poisson's ratio 79–80, 81t
Polygonal method 19–21, 20f, 20t
  chessboard grid 21, 23f
  construction of polygons by perpendicular bisectors (correct and incorrect) 21, 21f, 22f
  perpendicular bisectors 21, 21f
  square-net grid 21, 22f
Population standard deviation 9
  and control charts 32–33, 33f
Portals 5
Powder factor 280
Power (P) 6, 153. *See also* Horsepower
  available 245–246, 246f
Power factor (PF) 153, 158–160, 160t
  correction 158, 160–161, 161f, 161t
  correction capacitors 153
  correction with tangent (kVAR) table 160, 161t
  triangles 160–161, 161f
Power systems 145. *See also* Compressed-air power, Electrical power
Production planning 22–23
Project scheduling 26–27
Propel time factor 281, 283t
Pumps and pumping 121. *See also* Pipes
  dewatering 126–127
  efficiency equation 121
  frictional head 123, 124t
  gallons per minute 127
  method of horizontal slices 126–127
  multiple impellers 123–124
  point of intersection 123

pump characteristic curves 121, 122f
  in series 123–124, 126f
  speed change 123, 124–125
  static head 125–126
  station-duty application 125–126
  test facility 121, 122f
  total head 122–123

Q

Quality control 31–32
  control charts 32–33, 32f, 33f
  distribution range 32–33, 33f
  lower control limit 32
  upper control limit 32
Quantity of a fluid (or gas) flowing through
    a duct 7

R

Rail haulage. *See also* Locomotives
  acceleration 218–219
  curve resistance 218
  deceleration 218–219
  drawbar pull 217
  grade resistance 218
  rolling resistance 218
  suggested rail weights 215, 216t
  track layouts 215, 216t
  turnouts 215–217, 216f
Recovery ratio 6
Rectifiers 153
Reserve estimation 19
  polygonal (area-of-influence) method 19–22,
    20f, 20t, 21f, 22f
  sequence 19
Resistance (R) 154, 155, 158t, 159t
Retreat mining 5
Return 5
Rim pull 246, 246f
Rock strength and other properties 79, 81t
Rolling resistance 7
  rail haulage 218
  trucks 245, 245t
Roof bolts 86–87
  guidelines for selecting 87
  load per bolt 86–87
  yield capacity 87
Room-and-pillar mining, subsidence
    control 90–94, 93f
Root mean square method
  and drum hoists 200–202
  and Koepe hoists 202–204, 203f
  and locomotives 219–220
RPS. *See* Maximum (static) rope pull (RPS)
Rubber-tired haulage systems. *See* Trucks,
    Underground face-hauling vehicles
Rubbing surface 5

S

Safety and health. *See also* Mine Safety and
    Health Administration (MSHA), Ventilation
  airborne contaminants 28
  coal mine regulations 28, 29t, 30t
  incidence rate 28
  MSHA indices 27–28
  noise control 28–31, 31t
  severity measure 28
Safety factor
  strata control 82, 94
  wire ropes 195, 196t, 198, 199f
Sample variance 8–9
Scrapers
  calculating number of pusher tractors
    286–287
  crawler tractors 286
  cycle times (empty) 286, 287f
  cycle times (loaded) 286, 286f
  wheeled tractors 286
Service support system 1, 2f
Severity measure (SM) 28
Shovels
  dumping radius 284
  operating efficiency 280, 281t
  reach 284
  shovel pit without coal fender 285f
  swing factor 281, 283t
Skip 5
Skip load (SL) 197
Skip weight (SW) 197
Slope stability 94–95, 94f
Spacing
  between rows of holes (b) 274
  calculating 277, 279
  defined 5
  ratio 276
Specific gravity (s.g.) 6
Split 5
Staff planning 24, 24t–26t
Standard deviation 9
  and control charts 32–33, 33f
Standardized units 2, 3t–4t
Statistics 8–9
  variation 8–9
Strain 6
Strata control 1, 2f, 79
  boundary stress concentration around
    openings 79–80, 82f, 83f
  compressive stress concentrations around
    openings of various shapes 80, 84f
  longwall shield-type roof supports and
    capacity 88, 89f
  M (ratio of horizontal to vertical stress)
    79–80
  pillar design and strength equations 82–86
  rock and coal strength and other
    properties 79, 81t
  roof-span design 86–87
  safety factors 82
  slope-stability analysis 94–95, 94f
  stresses around mine openings 79–80, 80f
  and subsidence 88–94, 90f, 91f, 91t, 92f,
    92t, 93f
Stress 6
  compressive 79–80, 84f
  lateral 79

M (ratio of horizontal to vertical stress) 79–80
  and safety factor 82
  tangential 80
  working 86
  working shear 86
Subdrilling
  length (J) 274, 278, 279
  ratio 276
Subsidence 88–89
  control in room-and-pillar mining 90–94, 93*f*
  defined 5
  development curve 90, 90*f*
  hyperbolic tangent function 89
  prediction 90, 91*f*, 91*t*, 92*f*, 92*t*
Surface extraction 271. *See also* Blasting, Drilling, Excavation
Surface haulage vehicles. *See* Trucks
Surge capacitors 153
Swell factor 7, 281, 282*t*
Systeme International d'Unites (SI) 2, 3*t*–4*t*

T

Terminology 2–5
Threshold limit values (TLVs) 28
Total resistance (effective grade) 245
Tractive effort 7. *See also* Locomotives
Transformers 153
Transmission lines 153
Trip, defined 5
Trucks 241, 245
  adhesion 246
  approximate coefficients of traction 246, 247*t*
  available power 245–246, 246*f*
  brake performance curves (safe speed) 248, 249*f*–251*f*
  cycle time 247–248, 247*f*, 248*f*
  effective grade (total resistance) 245
  grade resistance 245
  gross vehicle weight (GVW) 245
  horsepower 245
  performance calculation 247–248
  rim pull 246, 246*f*
  rolling resistance factors 245, 245*t*
  typical travel time (empty) 247–248, 248*f*
  typical travel time (loaded) 247–248, 247*f*
Turnouts 215, 216*f*
  actual lead 217
  components 215, 216*f*
  formulas 217
  length of curved closure rail 217
  length of straight closure rail 217
  middle ordinate 217
  radius 217
  switch angle 217
  theoretical lead 217
  theoretical toe distance 217
  variables used to determine dimensions 215–217, 216*f*

U

Underground face-hauling vehicles 241, 245
  analyzing productivity potential of 241
  cable-reeled 241, 244
  change point 243, 243*f*, 244
  cycle-time analysis and formula 242–244
  elemental times 241–242
  LCT (loader cycle time) 242–243
  LT (loading time) 242–243
  and mine layout 242, 242*f*
  switch-in/switch-out 244
  untethered (battery- or diesel-powered) 241, 244–245
  WSC (loader waiting time) 242–243
Units of measure 2, 3*t*–4*t*
Upper control limit 32
US Bureau of Mines table of airway friction factors 52, 53*t*
US Department of Labor. *See* Mine Safety and Health Administration (MSHA)

V

Valves 121
Velocity 5
  and acceleration 6
  and displacement 6
Ventilation 51–52
  airflow principles 52, 52*t*
  airway friction factors 52, 53*t*, 54*t*
  ample airway capacity, 51
  fan adjustments and mine resistance 57–58, 57*f*
  fans and fan laws 56–57, 56*t*
  and fugitive air 51
  multiple parallel airways 54–55, 54*f*
  pressure loss 52–54
  pressures in inches of water gage 52
  resistance factor 52–54, 55
  split airflow (splits) 55–56, 55*f*
  system design process for new mines 58, 59*f*
Venturi meters 121
Voltage (V) 153
Voltage drop 155–158

W

Water control 121. *See also* Pipes, Pumps and pumping
Water gage 52
Wattmeters 121
Wire ropes 195. *See also* Hoists
  maximum (static) rope pull (RPS) 195–197
  safety factors 195, 196*t*, 198, 199*f*
  selecting 195–197
  size selection 197–198, 199*f*
  specifications for 6 × 7-class 195, 196*t*
  specifications for 6 × 19-class 195, 196*t*
Work 6
Working shear stress 86
Working stress 86

About the Author

Dr. Christopher J. Bise is Professor and Chair of the Mining Engineering and Industrial Health and Safety programs at The Pennsylvania State University and is the holder of the George H., Jr., and Anne B. Deike Chair. His areas of expertise and research interests include mine planning and design, occupational health and safety, production analysis, mine equipment maintenance, and mine management. He has authored numerous publications in these areas, including the first edition of *Mining Engineering Analysis;* he edited the SME book, *Coal Mining Technology: Theory and Practice;* and contributed to the SME *Mining Engineering Handbook* and *Mine Health and Safety Management.*

Dr. Bise holds three degrees in Mining Engineering, a B.S. from Virginia Tech, and an M.S. and Ph.D. from Penn State. He also received an M.H.S. in Environmental Health Engineering from Johns Hopkins in 1998.

Dr. Bise began his career at Penn State in 1974 as a research assistant. Prior to that, he was a resident engineer for two underground mines in eastern Ohio for the Consolidation Coal Company. Upon completion of his M.S. in 1976, he was appointed to the faculty as an instructor and, upon completion of his Ph.D. in 1980, progressed through the professorial ranks to his current position.

Dr. Bise has served his professional society, the Society for Mining, Metallurgy, and Exploration (SME), on over twenty committees, many as chair. He became the Coal Division Chair in 1995, and has also served as a member of the Board of Directors. He was named a Distinguished Member of SME's Pittsburgh Section in 1994, received the Coal Division's Distinguished Service Award in 2000, and was named a Distinguished Member of the Society in March 2001.

In addition to being a member of SME, Dr. Bise is a registered professional engineer in the Commonwealth of Pennsylvania, a member of the National Academy of Forensic Engineers, the Society of Professional Engineers, the American Conference of Governmental and Industrial Hygienists, and is a Certified Mine Safety Professional.

Dr. Bise has received the "Careers in Coal" Mining Engineering Professor award, Penn State's "Matthew J. and Anne C. Wilson Outstanding Teaching Award," the "PCMIA Stephen McCann Award for Educational Excellence," the "Old Timer's Award for Excellence in Mining Engineering Education," Virginia Tech's "Distinguished Mining Engineering Alumnus Award," SME 2002 Ivan B. Rahn Education Award, and Penn State's "Excellence in Academic Advising Award" in recognition of his achievements in mining-engineering education and advising.